ASPRS Annual Conference 2020

ISPRS Archives Volume XLIV-M-2-2020,2020

Online
22 – 26 June 2020

Editors:

K. Schuckman
T. Jordan

ISBN: 978-1-7138-2043-7

Printed from e-media with permission by:

Curran Associates, Inc.
57 Morehouse Lane
Red Hook, NY 12571

Some format issues inherent in the e-media version may also appear in this print version.

International Society for Photogrammetry and Remote Sensing (ISPRS)
c/o
Leibniz University Hannover
Institute of Photogrammetry and GeoInformation
Nienburger Str. 1
D-30167 Hannover, Germany

www.isprs.org

Additional copies of this publication are available from:

Curran Associates, Inc.
57 Morehouse Lane
Red Hook, NY 12571 USA
Phone: 845-758-0400
Fax: 845-758-2633
Email: curran@proceedings.com
Web: www.proceedings.com

TABLE OF CONTENTS

UAS LIDAR MAPPING OF AN ARCTIC TUNDRA WATERSHED: CHALLENGES AND OPPORTUNITIES

A. D. Collins[1], C. G. Andresen[1‡], L. M. Charsley-Groffman[1†], T. Cochran[2], J. Dann[1], E. Lathrop[1], G. J. Riemersma[2], E. M. Swanson[1], A. Tapadinhas[2], C. J. Wilson[1]

[1]Earth and Environmental Sciences Division, Los Alamos National Laboratory, Los Alamos, NM 87545, USA -(acollins, jdann, elathrop, emswanson, cjw)@lanl.gov;
[2]Routescene, Inc, 200 Troon Trail Durango, CO 81301, USA - tomcochran3@routescene.com, gjr@mapix.com, andre@mapix.com
[‡]Current Affiliation: Department of Geography, University of Wisconsin-Madison, Madison, WI 53706, USA - candresen@wisc.edu;
[†]Current Affiliation: Landpro Ltd., 13 Pinot Noir Drive, PO Box 302, Cromwell, Central Otago 9342, New Zealand - lauren@landpro.co.nz

KEY WORDS: Arctic, Best practices, Drone, LiDAR, UAS

ABSTRACT:

Uncrewed aircraft systems (UAS) are increasingly used across disciplines in academic research. We deployed a heavy-lift UAS (<25 kg) for research in the Arctic tundra, a remote and complex landscape. Conducting UAS work in this location required adapting our standard field approach to include both the unique challenges of working in these locations with those specific to UAS work. We collected metadata on each field campaign and analyzed our expended efforts and the contributors to our successes and failures. We formulated a set of best practices to address each challenge in a systematic way, addressing each with the underlying goals of maximizing system and team resilience, operational efficiency, and safety. By adopting a structured set of best practices tenets into our UAS work in the Arctic, we achieved greater project success and we recommend integrating such methods into similar projects of high importance or consequence, especially for UAS LiDAR work in the Arctic.

1. INTRODUCTION

Recent advancements in technology sectors including lithium batteries, microelectronics such as accelerometers, and multirotor software has allowed multiple manufacturers to develop and bring low cost uncrewed aircraft systems (UAS) options to market. As the UAS user base began to change from being dominated by hobbyists or highly-specified professionals to consumers, product offerings started to incorporate larger airframes and additional safety features that made adoption possible by industrial and research organizations, requiring the ability to deploy heavier payloads for multiple uses. Now, in parallel with increases in stability and endurance of UAS platforms, miniaturization is occurring in a wide variety of payloads that were traditionally borne on crewed aircraft, and instruments not previously used aboard aircraft are being developed and designed for UAS.

The technological improvements in UAS and associated sensors have gained popularity in many disciplines including surveying and mapping for a wide range of environmental applications (Hugenholtz et al. 2012; Bakuła et al., 2016; Jozkow et al., 2016; Sankey et al., 2017, 2018). However, the requirements for each use case differs depending on the complexity of the UAS system and its payload. For example, relatively simple operations such as taking a few airborne photos may involve operations with a small, technologically mature aircraft and an inexpensive payload, which may require a single pilot, no additional ground personnel or activities and little pre-planning. In contrast, more complex operations with larger, technologically less mature aircraft and costly payloads may require multiple pilots, visual observers and ground personnel, logistical planning, training, and the establishment of robust safety procedures.

Our study falls in the second aforementioned scenario of complex operations, where we examine the use of UAS to deploy a light detection and ranging (LiDAR) system to map vegetation and topography in the Arctic tundra of Alaska. The complexity of the UAS and LiDAR payload, in addition to the remoteness and extreme conditions of the study site, made these surveying campaigns very challenging. Given the lack of literature or other resources for best practices and experiences deploying complex UAS in difficult locations (e.g. Arctic), this manuscript seeks to help close that gap by relaying the knowledge and experiences we gained from our UAS operations.

2. MAIN BODY

2.1 Science in the Arctic

This use case describes UAS operations in the Arctic tundra of Alaska, USA as part of a US Department of Energy led, multi-national-laboratory and university project called the Next Generation Ecosystem Experiments (NGEE) Arctic. As permafrost systems in the arctic warm, the landscape is subject to rapid changes such as vegetation shifts and releases of landscape-stored CO^2 and methane into the atmosphere, all of which have potential impacts to natural systems across the globe (Schuur et al. 2008; Grosse et al. 2016). The overall goal of NGEE Arctic is to improve global climate model predictions through the better understanding and representation of the Arctic terrestrial system.

The ability of NGEE Arctic to perform scientific observations, experiments, and model earth systems processes depends upon using high-quality data products. Some such data are collected and available outside of the project such as satellite data products, but many of the science questions behind NGEE Arctic

https://doi.org/10.5194/isprs-archives-XLIV-M-2-2020-1-2020

require detailed surface data products at targeted locations and higher resolutions than is otherwise available. In addition, high resolution maps of topography and vegetation are needed at multiple time points to understand landform evolution and vegetation dynamics at multiple spatial and temporal scales given the transient and dynamic Arctic tundra system.

In order to obtain data products necessary to meet the project goals, we determined that mapping field locations with a LiDAR instrument aboard an uncrewed aerial vehicle (UAV) was the most effective solution. Although we considered using photogrammetric techniques to create the maps, LiDAR was selected to eliminate dependence on good ambient lighting conditions (which can be difficult to obtain at our locations), reduce the need for extensive ground control, and increase the potential data resolution at fine scales and through dense vegetation.

We tasked our field observation team with incorporating UAS into the regular schedule of seasonal fieldwork and identified a subset of the team to focus on details specific to UAS work. This UAS team underwent UAV and payload training, and assembled the documentation, protocols and approvals necessary for safe and efficient UAS work. Since the NGEE Arctic project began prior to including UAS operations, a rigorous fieldwork program was already in place. However, although additional complexities were expected during UAS work, many were difficult to anticipate, and we found no direct resource to guide our use of UAS in arctic tundra.

2.2 Field UAS Campaign

2.2.1 Field Location. In this paper, we examine two field campaigns over consecutive summers in 2017 and 2018 to map portions of a ~2.2 km watershed on the Seward Peninsula of Alaska, USA. The field location is along the Teller road, outside of Nome, AK at latitude 64°44'7.45"N and longitude 165°57'35.38"W (Fig 1). The field site consists of undulating tussock-tundra and shrubland vegetation with a S/SE aspect and 210 m elevation gradient.

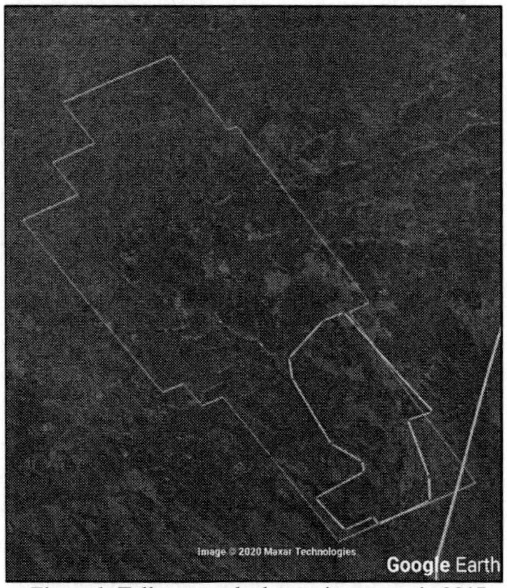

Figure 1. Teller watershed mapping extent in 2017 (green) and 2018 (yellow). Image footprint is approximately 3km by 2km. Teller road shown in white. Map data: Google, Maxar Technologies.

2.2.2 Extent. In 2017, the entirety of the watershed was mapped at a UAV forward speed of approximately 6 m/second at an altitude of 35 m above ground level (AGL), with the LiDAR instrument in single-return mode to obtain point cloud data with a density of 300-500 points per meter square. In 2018, we returned to the site and mapped only about 20% of the watershed, focusing on the densest areas of aboveground vegetation, and collected dual-return LiDAR data with point densities in the range of 5,000 points per meter square. In the 2018 campaign, we flew the UAV lower and slower, at 2 m/second forward speed and 20 m AGL.

2.2.3 Team. The UAS field time was comprised of four personnel in 2017 and five in 2018. The team included the same two UAS pilots and the same operations coordinator for both 2017 and 2018. The roles of payload operator and ground control surveyor were conducted by a single person in 2017, and split between two personnel in 2018.

2.2.4 Equipment. In 2017, we used a Vulcan Raven UAV (Vulcan UAV Ltd., Mitcheldean, Gloucestershire, UK) with a coaxial quadcopter design (Fig 2; Table 1). In 2018, we operated a DJI Matrice 600 Pro (Table 1; M600, Shenzhen DJI Sciences and Technologies Ltd., Shenzhen, Guangdong, China), which is a hexacopter platform. Both aircraft were powered with lithium-polymer batteries and electric motors, and controlled with DJI A3-based autopilot flight controllers. For all flight operations, waypoint missions were created and flown using UgCS software (SPH Engineering, Riga, Latvia) and flown in a boustrophedonic ("lawnmower") pattern with at least 50% side overlap.

Figure 2. Vulcan Raven aircraft carrying Routescene UAV LiDAR System LidarPod and GNSS antennas, and Sony α6000 payloads.

For both campaigns, we employed a Routescene UAV LiDAR System (Routescene, Edinburgh, Scotland, UK), which is comprised of an RTK GNSS telemetering ground unit, and a rotating 32-laser Velodyne scanner and RTK GNSS/INS air unit (Fig 2).

Ancillary equipment included portable generators, battery chargers, full duplex radio communication headsets, two-way radios, waterproof and shock-resistant rugged hard cases, differential GPS survey system, ground control targets, safety equipment, and multiple laptop computers and mobile devices. To provide a forgiving and even surface on which to conduct takeoffs and landings with the UAV, we built a 2.4 m by 2.4 m square landing pad. The landing pad was constructed with an underlying wooden dimensional lumber frame and top made of rigid foam board (R-TECH Insulation Panels, Insulfoam, Puyallup, WA, USA).

https://doi.org/10.5194/isprs-archives-XLIV-M-2-2020-1-2020

	2017 Vulcan Raven	2018 DJI M600	2019 Gryphon
Time per flight (minutes)	14	19	23
Maximum payload (kg)	3.9	5.5	12.3
Turnaround Time (minutes)	20	13	10
System Volume (m^3)	0.32	0.51	0.62

Table 1. Aircraft specifications for 2018 and 2018 field campaigns in Alaska. Details of an aircraft procured in 2019 shown for comparison. System volume includes rugged hard cases for airframe and batteries.

2.3 Campaign Challenges

2.3.1 Terrain at the field site is comprised of large hillslopes with undulating topography mostly covered by graminoid tundra, including tussocks formed by tussock cottongrass (*Eriophorum vaginatum* L.), dwarf shrubs, mosses and lichens (Miller et al. 1984). Generally, tussock tundra may reach heights over 50 cm. Intertussock spaces may be inundated with standing or flowing water during the summer, and small (<1 m) patches of exposed bedrock are occasionally present. In areas suitable for growth of large shrubs, patches of densely distributed alder and willow over 2 m in height exist. Areas of bare soil were rarely encountered.

Overland travel by foot was challenging due to uneven distribution and steep slope angles of tussocks, and required sturdy waterproof footwear to prevent cold exposure and injury. Additionally, walking on vegetation could be slippery, the amount of vegetation compression underfoot varied, and views of foot placement were often obscured due to presence of vegetation over ankle height. Locations of standing water were difficult to predict and each footstep required attention and focus to reduce injury potential, though no major injuries were sustained.

Transport of personnel and equipment was only possible by foot, as motorized travel could impact the sensitive ecosystem at the center of our research. We hauled equipment by hand, carrying backpacks, pulling sleds across tundra, (designed for use in snow, though no snow was encountered), and wheeling a two-person game cart (GameTote, Fort Collins, CO, USA). We were able to drive near the site at the southernmost boundary on an improved road and accessed locations within the watershed by either hiking up an unimproved road near the eastern boundary that extended to the north of the site, hiking directly across tundra, or some combination of hiking unimproved road and tundra.

2.3.2 2017 Weather was variable during the 12-day August campaign, with daytime temperatures between 3°C and 7°C for all but one day, which reached over 17°C. Rain was common with about half of flight days including rain showers during some point in a given day. Similarly, a little over half of the field days were windy, with wind speeds between 5 m/s and 11 m/s. Low temperatures contributed to slow foot travel, lower carrying capacity by personnel due to the need to carry and wear additional cold-weather gear, and reduced fine motor dexterity and thus slower equipment setup, operation and takedown. Turnaround time between flights (Table 1) was likely lowered, though it is difficult to determine how much was due to cold weather. Low temperatures did not significantly impact equipment, as electronics including batteries, screens, motors, etc. performed within their operational limits. Wind speeds did approach our maximum UAS takeoff and landing wind tolerance of 12 m/s, but conditions were generally favorable for flight. Rain did reduce our operational efficiency, as infrequent showers would cause a pause in operations and protection of equipment, initially with tarps, and later with a six-person camp tent. Transport of equipment was not affected by water, as equipment was stored in waterproof protective hard cases or tarped.

2.3.3 2017 Equipment reliability proved a source of delays and lost data for numerous reasons. Foremost, the Vulcan Raven airframe experienced such pre-campaign vibration during training and testing at our home elevation of 2100 m in Los Alamos, NM, that propeller bolts broke, propellers were ejected and motor mount bolts were fractured. After significant troubleshooting and testing, we changed from tri-blade folding carbon propellers to fixed dual-blade propellers and instituted a rigorous airframe check and bolt check procedure between flights. Since the airframe was configured for operations near sea level, we expected reduced vibration and resonance issues in Alaska. Although the Vulcan Raven did perform better in Alaska with more thrust and yaw authority, during one flight approximately 600 m away from the pilot, the airframe turned around and flew backward toward the set home point and was unresponsive to pilot input. About 100 m laterally before reaching the home point, the UAV descended faster than a normal landing approach, at an estimated 5 m/s and crash landed on soft tussock vegetation. Upon examination, we determined that vibration resulted in detachment of a MT60 bullet connector between a rear motor and electronic speed controller and the aircraft lost the use of that motor during flight. Given the lift redundancy of the other seven rotors, the aircraft maintained airborne status until landing. We also discovered breakage of a motor plate standoff bolt, though it is difficult to determine whether the bolt broke during flight or as a result of the hard landing. The airframe and payload was largely undamaged and after 1.5 days, we were able to make sufficient repairs and test for airworthiness to return the entire system to service and complete the campaign.

Additional transient equipment malfunctions occasionally further delayed operations. At times, both the UgCS waypoint software and DJI Go software suffered periods of unresponsiveness. On two occasions, the onboard-aircraft DJI A3 flight controller settings reverted to factory settings and it was necessary to reconfigure the flight controller prior to additional operations. Wireless connection (over 2.4 GHz wifi) between mobile device or laptop to handheld DJI Lightbridge 2 controllers and Routescene Ground Station intermittently failed to connect and toggling airplane mode, or resetting devices was sometimes necessary. The Routescene UAV LiDAR System required infrequent system restarts due to inertial navigation system (INS) errors and once required repositioning of the infrared camera trigger used to capture photogrammetry during LiDAR data collection.

2.3.4 2018 Weather. The ~two day field campaign in July 2018 saw more favorable weather with days near 15 °C, no rain and light ~3 m/s wind, though the rest of the campaign at other nearby field locations (not covered in this paper) included constant heavy rain for two of 13 days.

2.3.5 2018 Equipment Reliability. The M600 UAS platform was generally reliable, although twice the DJI software displayed electronic speed controller errors and system restarts were necessary to eliminate the errors. We had the use of four complete sets of batteries and more than once, flight operations were slightly delayed to await battery charging, but could have been alleviated with five complete battery sets. Similar software unresponsiveness, glitches, and freezes occurred with UgCS and DJI platforms as in 2017, as did occasional wifi connection issues. The Routescene UAV LiDAR System required a system reboot once or twice, but connectivity between air unit and ground unit, and between mobile device and ground unit was satisfactory and QA/QC criteria during data collection was met without incident.

2.4 Metadata and Analytical Techniques

2.4.1 Metadata Content and Analysis. This paper presents metadata associated with each of our field campaigns. The entirety of the 15-day 2017 field campaign was spent LiDAR mapping the Teller site, while two of 13 days in 2018 in the field were spent collecting LiDAR data at the same field site. For the purposes of this analysis, we chose not to include the remaining 11 days in 2018 mapping other sites due to different daily travel times, local weather patterns, access points, and terrains.

To determine the time spent in person hours for different activities, the number of hours for each activity category (for example, "active flight operations") was summed for all personnel. The percentage contribution of that activity was then calculated by dividing the category sum by the total number of hours spent on the campaign. For example, in 2017 the team spent a total of 68 person hours on active flight operations, and the total campaign took 967 person hours, so 68/967 = 7% of the campaign effort that was spent actively flying the UAS.

In 2017, the raw number of hours spent on each activity category was used to calculate the percentage of the total campaign effort. In 2018, the raw number of hours was used with two exceptions to prevent undue influence of time spent on the entirety of the campaign not specific to the two days at the Teller site: 1) the preparation and travel time was scaled by multiplying the number of hours spent in each category by 2/13 to represent the two days spent in the field, divided by the number of total field days; and 2) despite good weather at the Teller site in 2018, weather delays were included by multiplying the number of hours of total delays by 2/13 since we had the opportunity to choose good-weather days at Teller and to account for the random factor of weather in a multi-week field campaign. The total number of hours lost due to a completely rained-out field day was calculated by multiplying the number of personnel by the average field day length in 2017 (13.15 hours/day). Over both years, time spent in a given category was only counted if it pertained directly to the campaign of interest, e.g., obtaining UAS pilot licenses, which is a preparation activity that applies to all campaigns and not specifically to these campaigns, was not included.

Data sources presented herein came from multiple sources. The primary systematic sources were flight logs, preflight checklists, and battery logs and were used to determine the timing of each component of operations and corresponding levels of effort. Flight logs contained information including timing and location of each flight, environmental conditions, personnel, and flight and data characteristics. To ensure safe charging and to track battery use and performance, we recorded metrics on battery discharging and charging such as voltage at discharge and charge and the quantity of charge (ampere hours) for each charge cycle. In 2017, one team member kept a detailed field notebook that recorded general team efforts, notes on flight successes or failures, in-field arrival and departure times, temperature and weather, and other activity-related notes. Data from 2017 relied heavily on this field notebook data source. Additionally, other sources were sometimes necessary to constrain activity efforts and included emails, calendars, mobile device communications between team members and safety points-of-contact, FAA filings (notice to airmen, NOTAM) and LiDAR data files.

2.5 Results

2.5.1 Campaign Totals. In 2017, the UAS team mapped the entirety of the nearly 220 ha watershed, whereas the focus of the 2018 campaign was on less than a 50 ha targeted area (Table 2). Despite this, additional raw data was collected in 2018 as a result of flying lower, slower, and enabling the higher data collection rate of dual-return LiDAR. We averaged less than three flights per day in 2017, but more than doubled that in 2018. Changes and additions to the amount of equipment (including protective hard cases), resulted in a slight 9% increase in total equipment volume.

	2017	2018
Hectares Surveyed	216.52	45.65
Gigabytes of Data	45.5	50.2
# Flights	34	12
Flight Hours	6.52	3.84
Flight parameters	6m/s, 35 AGL	2m/s, 20 AGL
# Campaign Days	12	1.7
# Flights / Day	2.8	7.1
Total Person Hours	967	158
# Personnel	4	5
Equipment volume	1.8 m^3	1.96 m^3

Table 2. Campaign sums for 2017 and 2018. Note: equipment volume includes all UAS-related equipment but excludes personal gear.

2.5.2 Levels of Effort. As a percentage of the campaign in 2017, the largest effort was spent in preparatory activities, with a ~10% decrease in 2018 (Fig 3). The category of "Preparation" included gaining airframe flight currency immediately prior to departure, inventorying, packing and shipping equipment, and planning field logistics. Note that despite the great amount of time spent troubleshooting and testing the Vulcan Raven airframe before the 2017 campaign, these hours were not included in this analysis.

Travel to and from AK across both years was similar, and the relative amount of time spent driving between Nome and the field location increased in 2018, as shorter days were spent in the field. Hiking by foot to and from different staging areas within the watershed comprised 5% of the campaign in 2017, while hiking was unnecessary in 2018 due to close proximity to the staging area from the road. Ground control efforts including placement of ground control targets and DGPS survey of those targets took 5% of the campaign effort in 2017, and was reduced to half that in 2018 since some ground control targets were reused from the previous year.

https://doi.org/10.5194/isprs-archives-XLIV-M-2-2020-1-2020

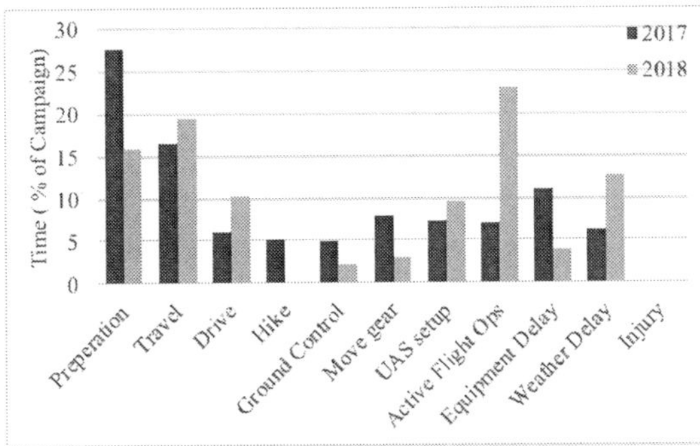

Figure 3. Time spent in each category as a percentage of the total campaign time.

The activities surrounding establishing UAS equipment on the ground and conducting flight operations varied by year. The team spent just 3.2% of campaign time physically moving gear by hand, sled or backpack in 2018 between the single staging area and nearby vehicles compared with 7.9% in 2017 among five staging areas across the watershed. UAS setup and takedown time was similar between years but active flight operations comprised only 7% of total campaign time in 2017 vs. nearly 23% in 2018.

Equipment delays were substantial in 2017, and at over 11% of campaign time, were the single largest allocation of on-the ground effort. Delays due to equipment issues remained in 2018 but were reduced to 4%. Weather (primarily rain) also impacted operations across both years and accounted for 6-13% of total time. No injuries to personnel were incurred either year.

Comparing year-to-year efficiencies, it took more effort in 2017 to fly the UAS for a given time period, and thus a given spatial area than in 2018 (Table 3). However, flying lower and slower in 2018 resulted in fewer hectares covered per flight hour.

	Person hours per flight hour		Person hours per hectare surveyed		Hectares surveyed per flight hour	
	2017	2018	2017	2018	2017	2018
Including Weather and Equipment Delays	14.3	4.4	4.5	3.4	0.9	0.7
Excluding Weather and Equipment Delays	11.9	3.6	3.7	2.9	3.2	1.3

Table 3. Levels of effort required to conduct one flight hour, map one hectare of land, and area surveyed per hour.

2.5.3 Data Quality was high both years with LiDAR returns from vegetation subcanopy features and ground. Obtaining a greater number of ground returns through areas of dense vegetation was a primary goal of 2018, and that goal was achieved (Fig 4).

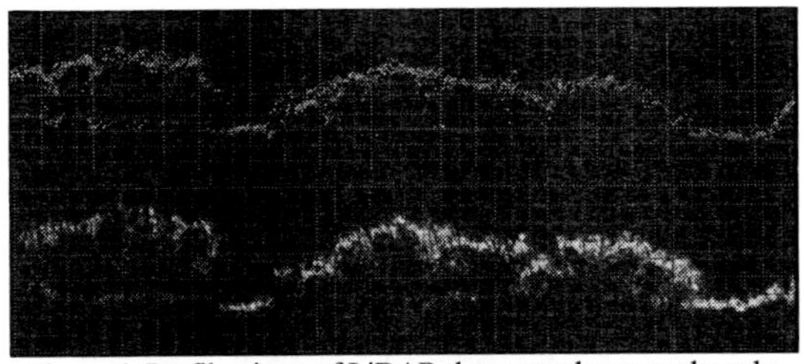

Figure 4. Profile views of LiDAR data over the same densely vegetated area in 2017 (top) and 2018 (bottom).

When comparing a digital terrain model (DTM) of a targeted area of tall shrubs between years, 2018 showed a large number of lower points, which may be ground or subcanopy vegetation returns compared to 2017 (Fig 5). Confidence was gained in this interpretation, as adjacent areas devoid of taller vegetation features registered identical (zero difference) values between years.

Figure 5. DTM of 2017 subtracted from 2018 of a targeted area. Yellows, oranges and reds indicate negative values (lower points in 2018), grey represents zero difference and blue represents positive values.

3. CONCLUSIONS

3.1 Best Practices Framework

We developed a UAS operational best practices framework for our work in the Arctic tundra through our preparatory activities, and by analyzing these data over the course of the two field campaigns. We arrived at a best practices framework by examining the challenges we faced and identifying opportunities for our team to respond differently and improve the success of the project.

We specifically examined the challenges presented by weather, equipment-related issues, terrain complexity, and vegetation density. We investigated the causes of impacts to our operations due to low temperatures, rain, and high winds. To identify opportunities to respond to equipment delays and failures, we looked to pinpoint system interdependencies and single points of failure. Terrain complexity increased foot travel time across the tussock tundra and led to additional safety risk, and we sought to minimize these impacts. The growth and distribution patterns of shrub patches at our field location made it difficult to obtain LiDAR returns beneath vegetation, and we examined our data to find areas where coverage was poor or error rates were high.

The best practices framework we developed are underlined by three basic principles that we apply to all challenges to maximize overall success: 1) increase system resilience; 2) increase operational efficiency; 3) increase safety. These three precepts are often interrelated and their application often begets additional benefits in multiple areas including team communication, inclusivity and performance satisfaction. In many cases, we arrived at meeting these principles by thinking about how we could either avoid or tolerate each individual challenge. Although our focus is in succeeding specifically in the Arctic tundra system in which we work, these practices can be generalized to other operations and for other systems. In the following, we examine each challenge and the steps we took to maximize resilience, efficiency and safety across our equipment and team, and achieve operational success.

3.2 Challenge: Weather

3.2.1 Precipitation. The main impacts of weather on our operations involved direct impacts (realized or potential) to our equipment and personnel (though indirect effects were incurred at times when driving was slowed due to heavy rain and corresponding road conditions). Since we worked with many electronic components like laptops, chargers, and the UAV itself, we identified avoidance as the best method to increase our three underlying principles of resilience, efficiency and safety. At first, we felt that utilizing tarpaulins and waterproof equipment cases would be adequate, and although resilience and safety criteria were largely met, the systems were not efficient. We later employed the use of a camping tent large enough to house our completely setup UAV (which takes time to dismantle and store in waterproof containers) and a 5-person team. The use of the tent met the two criteria and also greatly increased efficiency in taking shelter between intermittent rainstorms. Additionally, when possible, we used water-proof or -resistant equipment and clothing to tolerate some exposure to rain without expending effort in actively providing in-field protection.

3.2.2 Temperature. Low temperatures were not low enough to significantly impact our equipment functionality, but did affect personnel. Cold weather reduced hand and foot dexterity and could lower efficiency and expose the team to additional risk. We avoided cold weather by actively heating ourselves, using hand and foot warmers. Similar techniques could be used to heat equipment and batteries if necessary using items like hot water packs or electric heaters. Tolerating the cold by maintaining warmth was also identified as a necessary step through the use of insulated clothing for personnel or insulated containers for equipment. Utilizing a thermal mass such as water-filled containers could also be used to buffer impacts to equipment.

3.2.3 Wind primarily made UAV takeoffs and landings more challenging to conduct safely and efficiently, but also made verbal communication difficult. Avoiding wind during takeoffs and landings was sometimes achieved by careful positioning of the landing area if nearby land or vegetation features were available to offer wind protection, but often the wind was steady and we thus placed the landing pad in open areas to reduce turbulence from close objects. Using a large enough landing pad size was also important in providing flexibility for the pilot to land the UAV windy conditions. The landing pad was approximately four times the size necessary to land the size of our particular UAVs by our qualified pilots in ideal conditions. To improve safety for all team members during landing, especially during windy conditions, the pilot was positioned downwind of the landing area with the front of the aircraft facing away from the pilot such that the pilot could apply even pitch command while the UAV descended. Furthermore, any gusts that pushed the aircraft toward the pilot would ideally elicit a natural "push away" thumb response by the pilot to pitch the aircraft forward and away from the pilot. However, this practice can be challenged by the event of the aircraft being unresponsive and drifting towards the pilot. Team members were positioned behind the pilot to eliminate the need for the pilot to maintain spatial awareness of other personnel. Verbal communication was improved by adopting clear and concise communication styles similar to that used in aviation. We also used wireless headsets for easy hands-free communication when in close proximity to one another (~ 100 m) and handheld two-way radios for communication at farther distances (up to 2 km).

With multiple weather-related challenges, efficiencies could be reduced to zero despite all efforts if the weather was poor enough. Thus, planning the field campaigns for suitable weather, when possible, was helpful. More importantly, ensuring that the field team was prepared and rested enough to take advantage of periodic windows of good weather was critical. In the summer at high latitudes, with day lengths approaching 24 hours, we were able to occasionally work ~18 hour days to maximize data collection specially during good weather windows in this extreme environment.

3.3 Challenge: Equipment

3.3.1 Single points of failure, or system components which, if failed, would cease operations of the whole system, were examined in our equipment, and also as those failure points pertain to personnel, in their specialized roles or knowledge of equipment. Single points of equipment failure were critical particularly by the remoteness of the campaign where no specialized stores are available and shipping equipment in can take several days to weeks. Therefore, hardware component points of failure were identified and where possible, avoided. For example, both UAVs employed redundant motor systems such that a safe landing would be possible in the event of motor failure. Other critical equipment components were duplicated if budget allowed, such as battery chargers, mobile devices and tools. Avoiding single-point failures was also achieved by ensuring that multiple personnel were trained or had familiarity with critical components of the equipment, such that operations were possible if a single personnel was unavailable for some reason.

Often, failure points could not be avoided, such as with expensive items, or with those impractical to duplicate. In many cases, when we invested in quality equipment that was selected for durability and longevity, this minimized reductions in operational problems. We discovered that not all of the many hardware constituents of a UAS and associated operations are held to the same manufacturing standard and level of reliability, and that using cheaply or poorly made products often introduced great decreases in efficiency and thus increased mission costs.

3.3.2 Maintenance and testing of our equipment was critical to ensure safe and successful operations. We developed a rigorous system of checks and balances that we applied to the planning stages through campaign completion. Broadly, we enacted multiple checklists including packing lists, to-do lists, beginning-of-day checklists, pre-flight checks, post-flight checks, data quality checks, and multiple procedural documents and checklists specific to sub-systems and components. We designated certain parts of some checklists as necessary for double-checking by more than one individual, especially those related to flight-critical UAV components. On the daily or sub-daily timescale, any maintenance issue encountered was immediately corrected or noted, discussed, and decisions were made to continue or suspend operations pending correction of the issue. Over longer periods of time, flight hours and battery charge cycles were carefully tracked, and predetermined maintenance schedules for various components were adhered to.

Testing the multiple electronics systems was also important because functionality of so many different pieces of the entire system was necessary, often simultaneously, to complete a given flight. We identified interdependencies of various systems, for example a wireless router through which numerous devices communicated and tested those systems in different physical and software configurations. Testing the systems often revealed single points of failure or recurrent points of weakness and gave us the opportunity to increase resilience or efficiency with redundant or reconfigured equipment. Testing in assorted arrangements also provided a springboard for developing contingency plans when failures or difficulties were encountered. In order to test systems such as satellite reception of the UAV at high latitudes, along with mission planning software functionality and accuracy of the base digital elevation maps it relied upon, we often deployed an inexpensive small DJI Phantom 4 Advanced UAV, which had similar or identical software to the large and costly primary UAV and payload before it was deployed.

Often in concert with testing various systems prior to campaigns, maintaining personnel training was helpful in keeping the team prepared for safe and efficient operation of all equipment and software. Though pilot flight currency with the UAV in question was the top priority in maintaining training, currency of the payload operator with the payload system, the surveyor with the survey equipment, the operations coordinator with the planning software and team UAS currency in general allowed for smooth operations. Since our team did not conduct UAS projects year-round, it was especially important to refresh our knowledge and skills before each campaign.

A surprisingly large part of maintaining a properly functioning UAS was related to keeping various batteries charged for use during daily operations. In addition to the UAV batteries, rechargeable batteries were present in or needed for items such as laptops, tablets, headsets, two-way radios, UAV controllers, UAV ground station, DGPS rover, DGPS controller and DGPS ground station, UAV LiDAR system ground station, handheld GPS, and small UAV batteries and controller. In all, 25 small (<100 watt hour) lithium batteries that were not integrated into equipment were utilized. All batteries were required on a daily basis and we employed scheduled charging at night among all team members.

3.3.3 Team communication was integral to our ability to operate equipment safely and effectively. We encouraged a communication style of clear and concise verbal and nonverbal communications, where background noise and misinterpreted messages could have serious consequences. We instituted a procedure of engaging in predefined conversations during operations in a sequential manner, identifying both the content of the interactions and parties involved. For example, immediately prior to takeoff, the operations coordinator would request verbal responses from multiple parties, starting with the pilot then the payload operator to confirm readiness of the UAV and LiDAR systems, after which the pilot would ask the backup pilot if all was clear for takeoff and upon confirmation, the pilot would notify the entire team of impending propeller spin up and takeoff.

We also fostered both a culture of accepting personal responsibility for actions, while recognizing that any one member of the team is fallible and that many parts of our procedures require a second or third person to verify that actions were performed correctly. Frequently throughout an operations day, we would create opportunities for the entire team to pause their activities to focus on discussing questions and concerns about previous or upcoming work. We monitored the levels of physical and mental fatigue by checking in with one another, and each team member was empowered with the ability to pause or stop the activities of the team for any reason.

3.4 Challenge: Terrain

The tussock-tundra terrain was challenging primarily because foot travel was physically demanding and slow. Since we weren't able to modify the terrain, we raised resilience, efficiency and safety by evaluating multiple methods of travel such as by foot, automobile, helicopter or all-terrain vehicle. Despite exploring motorized options, we deemed foot travel to be the best option and considered other ways to haul equipment. We were hopeful that a game cart, designed for hauling big game out of complex terrain would be effective, but instead, dragging sleds across the low vegetation proved to be the best choice, as each team member could work at their own pace, and transport via sled was most efficient.

We aimed to maximize the safety of our team in a few ways. First, team members were asked to evaluate their own physical fitness with respect to hiking many kilometers in such terrain. We then identified and used appropriate personal equipment, namely sturdy waterproof footwear, to protect against ankle sprains and wet feet. All team members were required to undertake safety-related training such as bear safety and first aid training, and to develop and follow emergency procedures. As an additional safety measure, the team carried satellite phones in the event of emergency.

3.5 Challenge: Vegetation Density

The presence of dense vegetation presented unique challenges at our field locations. Poor resolution of the ground surface beneath shrubs resulted in difficulty preparing accurate DTMs to meet project goals. To combat this problem, we flew lower to the ground and with slower forward speed to increase the number of returns per square meter. We also learned to fly the LiDAR instrument at oblique angles to the ground to improve the chances of penetrating vegetation.

We observed high error and reduced accuracy in multiple areas of our data collection footprint and identified multiple paths to making improvements. First, we constrained the process of INS initialization of the LiDAR instrument by flying the UAV in smooth movements similar to fixed wing aircraft flight, with gentle throttle inputs, and with small variations about all axes of pitch, roll and yaw. Second, we considered the satellite configuration that would be present at the time of flights to ensure that the UAV LiDAR system would have adequate coverage to obtain accurate RTK solutions. Lastly, we deployed additional LiDAR-reflective ground control and maximized the survey accuracy of those points to improve ground validation.

3.6 Summary

Our operational success in UAS LiDAR work in the arctic tundra relied greatly on the efforts of the entire field team and LiDAR vendor to develop a set of best practices specific to our needs. We built our UAS program around three fundamental concepts: resilience, efficiency, and safety. Improvements in system and team resilience was achieved through incorporating adaptability

This contribution has been peer-reviewed.
https://doi.org/10.5194/isprs-archives-XLIV-M-2-2020-1-2020 | © Authors 2020. CC BY 4.0 License.

with redundant and high-quality equipment, and well-trained communicative personnel. Efficiencies were realized by maximizing our return on effort and examining all aspects of operation. Safety was integrated into the underlying operational framework of the project and facilitated every other component of our work.

We accomplished our UAS project goals by collecting metadata from field logs, communications, and documentation, and leveraged that data to make improvements. Analyzing this metadata revealed inefficiencies and showed us where we spent the greatest amount of effort and funds. Based on these analyses, we were able to modify our strategies and achieve better results on subsequent campaigns. Although we did not seek to include metadata collection and analysis from the outset of our UAS program development, we now recognize the immense value of such tools to augment our research goals and recommend these approaches for future projects.

ACKNOWLEDGEMENTS

Funding was provided by the Next-Generation Ecosystem Experiments (NGEE Arctic) project, supported by the Office of Biological and Environmental Research in the U.S. DOE Office of Science.

REFERENCES

Bakuła, K., Ostrowski, W., Szender, M., Plutecki, W., Salach, A., & Górski, K. (2016). Possibilities for using lidar and photogrammetric data obtained with an unmanned aerial vehicle for levee monitoring. *International Archives of the Photogrammetry, Remote Sensing & Spatial Information Sciences*, 41.

Conte, G., Kleiner, A., Rudol, P., Korwel, K., Wzorek, M., & Doherty, P. (2013). Performance evaluation of a light-weight multi-echo lidar for unmanned rotorcraft applications. *International Archives of the Photogrammetry, Remote Sensing & Spatial Information Sciences*, 40, 4-6.

Grosse, G., Goetz, S., McGuire, A. D., Romanovsky, V. E., & Schuur, E. A. (2016). Changing permafrost in a warming world and feedbacks to the Earth system. *Environmental Research Letters*, 11(4), 040201.

Hugenholtz, C. H., Moorman, B. J., Riddell, K., & Whitehead, K. (2012). Small unmanned aircraft systems for remote sensing and earth science research. *Eos, Transactions American Geophysical Union,* 93(25), 236-236.

Jozkow, G., Totha, C., & Grejner-Brzezinska, D. (2016). UAS TOPOGRAPHIC MAPPING WITH VELODYNE LiDAR SENSOR. *ISPRS Annals of Photogrammetry, Remote Sensing & Spatial Information Sciences*, 3(1).

Miller, P. C., Miller, P. M., Blake-Jacobson, M., Chapin, F. S., Everett, K. R., Hilbert, D. W., ... & Roberts, S. W. (1984). Plant-Soil Processes in Eriophorum Vaginatum Tussock Tundra in Alaska: A Systems Modeling Approach: Ecological Archives M054-002. *Ecological Monographs*, 54(4), 361-405.

Sankey, T., Donager, J., McVay, J., & Sankey, J. B. (2017). UAV lidar and hyperspectral fusion for forest monitoring in the southwestern USA. *Remote Sensing of Environment*, 195, 30-43.

Sankey, T. T., McVay, J., Swetnam, T. L., McClaran, M. P., Heilman, P., & Nichols, M. (2018). UAV hyperspectral and lidar data and their fusion for arid and semi-arid land vegetation monitoring. *Remote Sensing in Ecology and Conservation*, 4(1), 20-33.

Schuur, E. A., Bockheim, J., Canadell, J. G., Euskirchen, E., Field, C. B., Goryachkin, S. V., ... & Mazhitova, G. (2008). Vulnerability of permafrost carbon to climate change: Implications for the global carbon cycle. *BioScience*, 58(8), 701-714.

The International Archives of the Photogrammetry, Remote Sensing and Spatial Information Sciences, Volume XLIV-M-2-2020, 2020
ASPRS 2020 Annual Conference Virtual Technical Program, 22–26 June 2020

STUDY OF ACTIVE FARMLAND USE TO SUPPORT AGENT-BASED MODELING OF FOOD DESERTS

Saurabh S. Dhamankar[1], Leila Hashemi-Beni[1*], Lyubov A. Kurkalova[1], Chyi Lyi Liang[1], Timothy Mulrooney[2], Manoj Jha[1], Gregory Monty[1], Haoran Miao[1]

[1] North Carolina A&T State University
[2]North Carolina Central University
*lhashemibeni@ncat.edu

KEY WORDS: Food Desert, Agent-Based Modeling, Land-Use, Spatial Analysis, Temporal Analysis

ABSTRACT:

Food desert (FD) is the area that has limited access to affordable and nutritious foods such as fresh fruits, vegetables, and other healthful whole foods. FDs are important socio-economic problems in North Carolina (NC), potentially contributing to obesity in low-income areas. If farmland is available, local vegetable production could potentially help alleviate FDs. However, little is known about land use and land-use transitions (LUTs) in the vicinity of FDs. To fill this knowledge gap, we study the farmland use in three NC counties, Bladen, Guilford and, Rutherford, located in Coastal, Piedmont, and, Mountain regions of the state, respectively. The analysis combines the United States Department of Agriculture (USDA) 2015 FD/NFD delineation of census tracts, and geospatial soil productivity and 2008-2019 land cover data. The understanding of farmland use is expected to contribute to the development of LUT components of FD Agent-Based Models (ABM).

1. INTRODUCTION

The term "food desert" (FD) was formally introduced to describe areas with limited access to affordable nutritious foods, particularly areas in lower-income neighborhoods. The definition has led to the development of national and regional maps that focus efforts on equity in food access. Recognition of FDs also marks a strategic change in public health approach to obesity prevention (Karpyn et al., 2012). A growing body of research suggests that the suburbanization of food retailers has contributed to the emergence of urban FDs. While more and more large-format supermarkets are erected on suburban lands, smaller grocers in older central-city neighborhoods seem to be rapidly disappearing, leaving potential FDs in their wake (Larsen and Gilliland, 2008).

United States Department of Agriculture (USDA) has developed an web-based, national mapping tool that pinpoints the location of FDs at the census track level (https://www.ers.usda.gov/data-products/food-access-research-atlas/, accessed July 2020). The tool and the corresponding data have been used in multiple studies. For example, Smith et al. (2012) evaluated how recent trends and legislative developments have increased access to fresh and healthy food in New Jersey. In particular, since 2011, the state's legislative acts have allowed nonprofit corporations and associations located in select New Jersey cities to transform vacant properties into gardens for growing fruits and vegetables. The government has lowered the cost of urban agriculture projects by making public land available, tax-free, and thus contributing to alleviation of FD in the select cities. While Smith et al. (2012) focused on vacant urban properties as a source of land for expanding local vegetable production, potentially larger areas could come from farmland, the land use considered in this study.

In this paper, we investigate the use and the potential availability of farmland for vegetable production within three North Carolina (NC) counties, Bladen, Guilford, and, Rutherford. The study areas have been chosen to reflect the within-state geographic and biophysical diversity: the three counties are located in Coastal, Piedmont, and Mountain regions, respectively.

The exploration of the local food production approach to alleviation of FDs requires an understanding of not only land use at a specific point of time, but also land-use change. The change in land use/cover is one of the most profound human-induced alterations of the Earth's system. Land conversion for farming or other productive uses serves human well-being but also causes changes in ecosystem functions. Land use change emerges from the interactions among various components of the coupled human–landscape system, which then feeds back to the subsequent development of those interactions. Accrual of these short-term changes over time and the aggregation of these localized changes over space generate larger-scale emergent patterns of land use transition (LUT) and economic development (Le et al., 2008). Simulation models capturing changes in both space and time have a potential to explain the observed LUTs and to explore the possible future developments in land-use systems.

An ABM is a class of computational models for simulating the actions and interactions of autonomous agents to assess their effects on the system as a whole. An agent-based simulation is most commonly used to model individual decision- making and social and organizational behavior. In an ABM, a simulation environment is composed of agents and rules. Le et al. (2008) discusses the concept and theoretical specification of a multi-agent-based model for spatio-temporal simulation of a coupled human landscape system. The authors emphasize that the understanding of the LUT is imperative for understanding the rules of such coupled human landscape systems.

This contribution has been peer-reviewed.
https://doi.org/10.5194/isprs-archives-XLIV-M-2-2020-9-2020

The social, ecological, and economic nature of ABMs has a strong connection with Geographic Information Systems (GIS). The habitat of an agent and its surrounding environment influence the agent's decision-making, which in turn can influence or change the landscape the agent interacts with. For example, Swetnam et al. (2011) presents a GIS method to interpret qualitatively expressed socio-economic scenarios in quantitative map-based terms. The study builds scenarios using local stakeholders and experts to define how major land cover classes may change under different sets of conditions. The study formalizes spatially explicit rules of changes and presents future land cover maps, which can then be used to model ecosystem services. Such modeling of social agents' interactions over a period of the time, in a spatially referenced environment, provides a powerful tool for policymakers, planners, managers, and the public to understand the outcomes of the simulation.

To quantify the use and the potential availability of farmland for vegetable production, we overlay and analyze several data layers developed by USDA: geospatial soil productivity data, the 2008-2019 Crop Data Layer (CDL) land cover data, and the 2019 FD/NFD delineation of census tracts. We group all farmland in six broad categories, with the first two categories, vegetables & produce and berries & non-tree fruits corresponding to the land use that immediately produces healthful whole foods. The additional four categories correspond to the other farmland uses that could require alternative levels of investment to transition to the first two categories: fallow, hay, single season crops, and double season crops. For each of the county, we conduct spatio-temporal analyses to investigate whether FD and NFD areas differ in the relative importance of the six broad uses of farmland and farmland soil quality. We find significant between-county differences in the relative importance of the six broad uses of farmland. However, we find only limited differences between FD and NFD land use or soil quality within the counties.

This study contributes to the development and testing of LUT rules applied to agent-based economic models, with the ultimate goal of developing an ABM to assess the impacts of alternative FD policy interventions.

2. METHODOLOGY

2.1 Study Area and Data

The National Agricultural Statistical Service (NASS) of USDA groups counties in each state in Agricultural Statistical Districts by geography, climate, and cropping practices (USDA, 2000). To reflect the natural resources and farmland use diversity of NC, the study counties are selected from three different agricultural statistical districts (ASDs): Southern Coastal (Bladen county), Northern Piedmont (Guilford county), and Western Mountain (Rutherford county) (Figure 1).

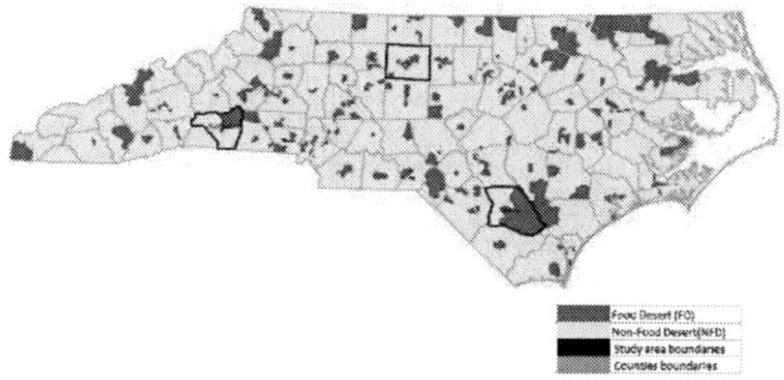

Figure 1. North Carolina food desert census tracts and study areas

Three data sets, all developed and maintained by the USDA, are used in the analysis. The location of FDs comes from the USDA Food Access Research Atlas (https://www.ers.usda.gov/data-products/food-access-research-atlas/, accessed July 2020), which provides food access indicators by census tract. We use the latest, 2015 data available, with the FD defined as low-income tract with at least 500 people, or 33 percent of the population, living more than 1 mile (urban areas) or more than 10 miles (rural areas) from the nearest supermarket, supercenter, or large grocery store.

To represent the soils' innate crop productivity, we use the National Commodity Crop Productivity Index (NCCPI) data layer coming from the 30m raster produced by the USDA National Resources Conservation Service (USDA/NRCS, 2020). The NCCPI ranks the inherent capability of soils to produce agricultural crops without irrigation on a scale of 0-1 with larger values indicating more productive soils.

The land use data come from the Crop Data Layer (CDL), an annual, raster, geo-referenced, crop-specific land cover data layer with the spatial resolution of 30m (USDA/NASS, 2020). The annual land cover classification includes broad categories such as urban/developed land, forests, and wetlands, and is specifically tailored toward crop identification (Lark et al., 2017). Approximately 70 land uses show in NC layers every year. We use all the yearly layers for which the CDL data are available consecutively for NC, 2008 – 2019.

The county data summaries (Table 1) reveal that Bladen has the largest, while Guilford has the lowest share of area in FD.

County	Area (Acres)	Food Desert Area (%)
Bladen	537,404	64
Guilford	397,284	08
Rutherford	342,238	31

Table 1: Summary statistics

2.2 Spatial Analysis

A spatio-temporal analysis has been carried out using ArcGIS by ESRI. The flow chart of spatial analysis is shown in Figure 2. At the initial step, Data Collection, the data are collected from various authentic sites. The FD data comes in a tabular (spreadsheet) format by the census tract. The census tract (CT) map has been downloaded from the U.S. Census Bureau web site (https://www.census.gov/geographies/reference-maps/2010/geo/2010-census-tract-maps.html, accessed July 2020). The FD map of NC was created using the spatial join of the CT map and the FD data.

The CDL is downloaded from the USDA/NASS (2020). The data for the state of NC has to be clipped out from the original downloaded map for data processing. For each of the yearly layers, 2008 to 2019, we consider only farmland categories (approximately 50 categories) that represent land uses that are either currently or relatively easily (within a year) could be devoted to production of vegetables and fresh produce. To reduce the total number of farmland categories to a manageable set, we reclassify farmland use in six broad FL categories (Table 2).

https://doi.org/10.5194/isprs-archives-XLIV-M-2-2020-9-2020 | © Authors 2020. CC BY 4.0 License.

FL category	FL category definition	Original CDL categories
Vegetables & produce (VP)	Vegetables and produce which are grown as annuals, or as short-time-span perennials	42-44, 46, 47, 49-54, 57, 206-208, 214-216, 219, 222, 227, 243-248
Berries & non-tree fruits (BF)	Berries and fruits which are grown as annuals, or as short-time-span perennials	48, 55, 209, 213, 221, 242, 250
Fallow (F)	Land that currently has no crops on it, but that could be brought back to production without any additional investment	61
Hay (H)	Grasses grown as a source of animal feed	37, 37, 176
Single season row crops (SC)	Row crops planted in the spring and harvested in the fall	1-25, 27-35, 38-41, 45, 56, 58-60, 205, 224, 229, 249
Double cropped (DC)	Two crops are grown during the same crop production year: one crop is planted in the fall and harvested in the spring, and the other crop planted in the spring and harvested in the fall	26, 225, 226, 230-241, 254

Table 2: Six farmland (FL) categories. Note: The complete list of the original CDL categories is available in USDA/NASS (2020).

The FL data is overlaid with FDP data to obtain the maps and the estimates of the areas. Then results were summaries in the form of charts.

The Soil Productivity Layer (SPL) is downloaded from the USDA/NRCS (2000) and is available in two parts for the state of NC. The two SPL maps were merged to cover the study areas in a single layer, NC SPL. The NC SPL is used in further analysis to estimate the average soil productivity of different areas and classes. The average productivity of farmland categories and/or areas is estimated using a zonal analysis applied to the corresponding categories and/or areas.

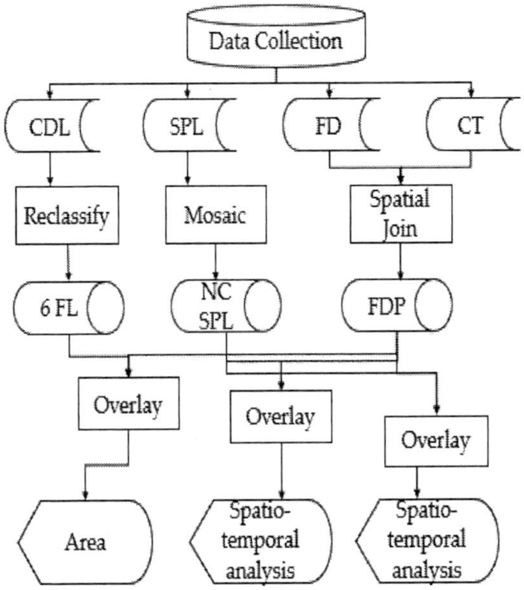

Figure 2. Flow chart of spatial analysis

3. RESULTS AND DISCUSSION

We find that the use of the FL is dominated by hay in the Rutherford County, and, to a smaller extent, in Guilford county, while single season crops dominate in Rutherford county (Figure 3 and Figure 4). These differences confirm to prior expectations because the western (Mountain) section of the state has a predictable weather patterns but a short growing season, and hence is not well suited for crop production. The length of the crop-growing season increases eastward, with the Coastal Plain being most suitable part of the state for growing crops (Brown and Vogt, 2015).

Overall, only a small share of farmland is currently under VP and BF, with the largest 2008-2019 average share recorded in Bladen County, barely above 0.2%, (Figure 5 and Figure 6). Because of these small shares, to better present the differences across counties, Figure 6 is focusing exclusively on VP. The climatic conditions suitable for crop production in general seem to be benefiting vegetable and fruit production as well: Bladen County has the highest VP and BF shares among the three study areas. The spatial pattern of farmland use in the county suggests that VP and BF uses are concentrated in only a few spots – the finding that warrants additional investigation.

The patterns of soil productivity (Figure 7) suggest that vegetable and fresh produce production favors a higher quality land in North Carolina: the average NCCPI of VP is approximately the same as that of SC, with both statistics being noticeably higher than those for the other four farmland uses.

The comparison of FD and NFD farmland use (i.e., Figures 3, 4 and 6) does not reveal any notable differences in farmland use inside versus outside of food deserts in the study areas. Nor there appears to be any difference in the average farmland soil productivity between FD and NFD (Figure 8).

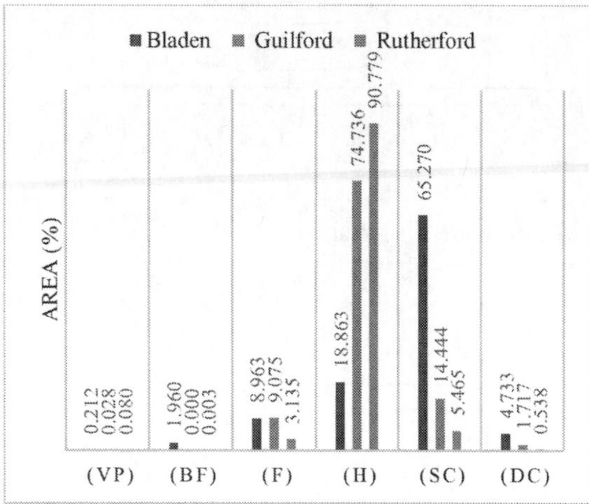

Figure 3. Food deserts: 2008-2019 average shares of the farmland area under alternative uses. Notes: VP = vegetables & produce, BF = berries & fruits, F = fallow, H = hay, SC = single season, DC = double season.

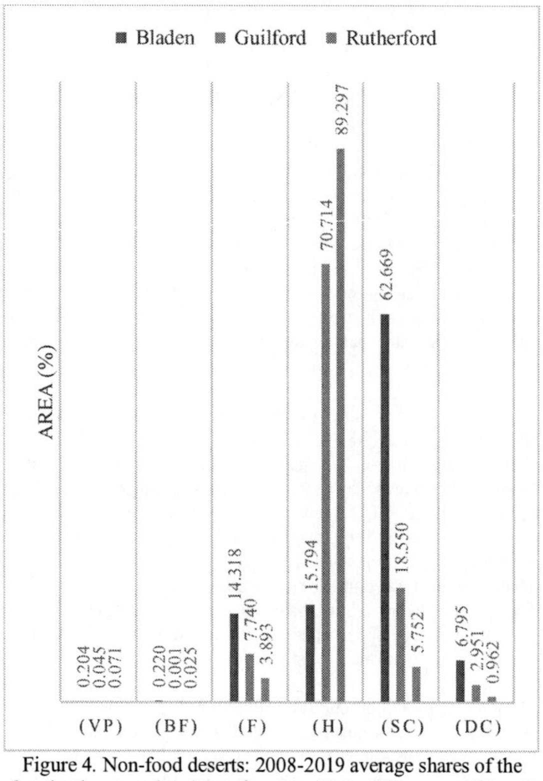

Figure 4. Non-food deserts: 2008-2019 average shares of the farmland area under alternative uses. Notes: VP = vegetables & produce, BF = berries & fruits, F = fallow, H = hay, SC = single season, DC = double season.

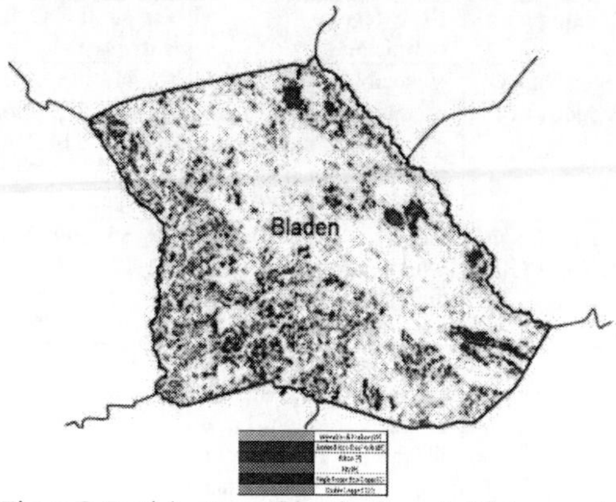

Figure 5. Spatial pattern of farmland uses in Bladen county, 2019.

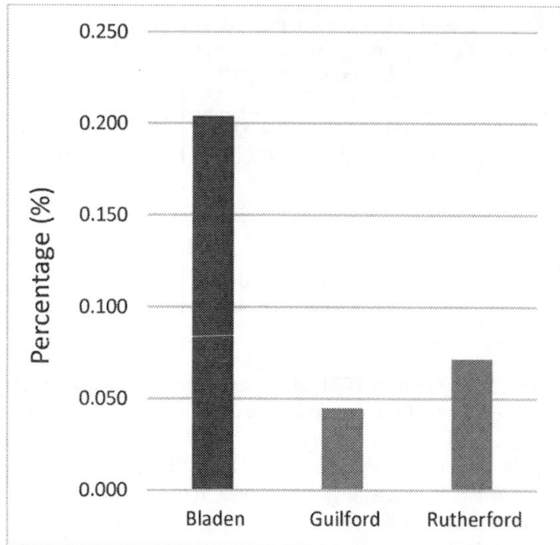

Figure 6. 2008-2019 average shares of the farmland area under VP

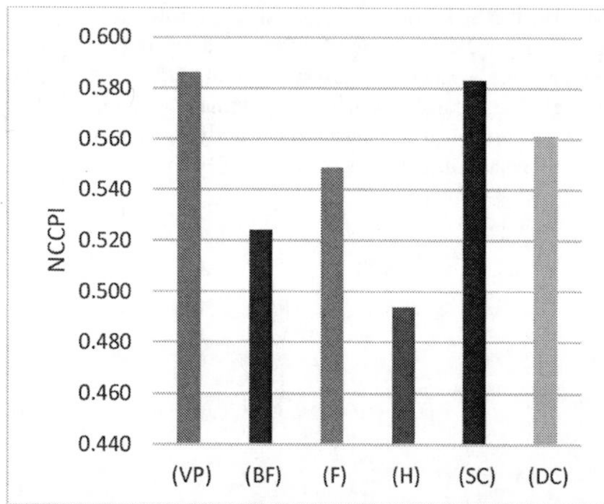

Figure 7. 2008-2019 average productivity of the farmland area under alternative uses. Notes: VP = vegetables & produce, BF = berries & fruits, F = fallow, H = hay, SC = single season, DC = double season.

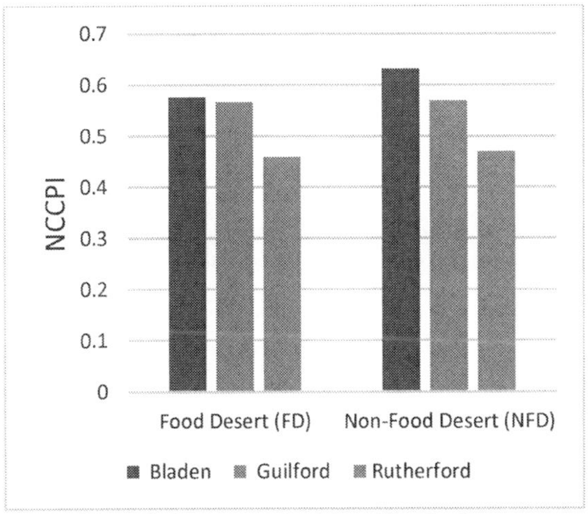

Figure 8. Average productivity of farmland area

4. CONCLUSIONS

Twelve years of CDL data, 2008 – 2019, were combined with soil productivity maps to analyze farmland use inside and outside of food deserts in three NC counties. We did not find any noticeable differences in the soil productivity or in the use of farmland between FD and NFD areas. The analysis revealed that highest productivity lands tend to be used to grow VP and SC, while the lower productivity land is double cropped, fallowed, used to grow hay, or berries and non-tree fruits.

The improved understanding of farmland use is expected to contribute to the development of LUT components of FD agent-based models. Our findings lead to two hypotheses to be investigated further. First, our results suggest that the expansion of VP production to fallow, hay, or double-cropped acreage may be limited by soil productivity. In consequence, the transition of farmland to vegetable production might require economic incentives that differ depending on the current land use. Second, there is an indication that vegetable and produce production may be clustered geographically. Future analyses would benefit from investigating which climatic and/or landscape factors contribute to specific geographic patterns of vegetable production in North Carolina.

ACKNOWLEDGEMENT

This material is based upon research work supported by the National Science Foundation (NSF) under grant No. 1824949.

REFERENCES

Brown, M.J.; Vogt, J.T. North Carolina's forests, 2013. U.S. Department of Agriculture, Southern Research Station, Resource Bulletin SRS-2015. 2015. Ashville, NC.

Karpyn, A., C. Young, and S. Weiss. "Reestablishing Healthy Food Retail: Changing the Landscape of Food Deserts." *Child Obes* 8.1 (2012): 28-30.

Lark, Tyler J., et al. "Measuring Land-Use and Land-Cover Change Using the U.S. Department of Agriculture's Cropland Data Layer: Cautions and Recommendations." *International Journal of Applied Earth Observation and Geoinformation* 62 (2017): 224-35.

Larsen, K., and J. Gilliland. "Mapping the Evolution of 'Food Deserts' in a Canadian City: Supermarket Accessibility in London, Ontario, 1961-2005." *Int J Health Geogr* 7 (2008): 16.

Le, Q. B., Park, S. J., Vlek, P. L. G., & Cremers, A. B. (2008). Land-Use Dynamic Simulator (LUDAS): A multi-agent system model for simulating Spatio-temporal dynamics of coupled human–landscape system. I. Structure and theoretical specification. *Ecological Informatics, 3*(2), 135-153.

Smith, Jim. "Encouraging the Growth of Urban Agriculture in Trenton and Newark through Amendments to the Zoning Codes: A Proven Approach to Addressing the Persistence of Food Deserts." *Vermont Journal of Environmental Law* 14 (2012): 71 - 100.

Swetnam, R. D., et al. "Mapping Socio-Economic Scenarios of Land Cover Change: A Gis Method to Enable Ecosystem Service Modelling." *J Environ Manage* 92.3 (2011): 563-74.

Tsai, Y., Zia, A., Koliba, C., Bucini, G., Guilbert, J., & Beckage, B. (2015). An interactive land-use transition agent-based model (ILUTABM): Endogenizing human-environment interactions in the Western Missisquoi Watershed. *Land Use Policy, 49*, 161-176. doi:https://doi.org/10.1016/j.landusepol.2015.07.008

U.S. Department of Agriculture (USDA). (2000). Farm Resource Regions. USDA Economic Resource Service Agricultural Information Bulletin Number 760. https://www.ers.usda.gov/publications/pub-details/?pubid=42299

U.S. Department of Agriculture, National Agricultural Statistics Service (USDA/NASS) Cropland Data Layer. {2020}. Published crop-specific data layer [Online]. Available at https://nassgeodata.gmu.edu/CropScape/ (accessed {May 5, 2018}; verified {February 4, 2020}). USDA-NASS, Washington, DC.

U.S. Department of Agriculture, National Resources Conservation Service (USDA/NRCS). (2000). USA Soils Crop Production [Layer package]. ESRI on ArcGIS Online, URL: https://landscape11.arcgis.com/arcgis/ [July 2020].

Wang, Tong, et al. "Determinants of Motives for Land Use Decisions at the Margins of the Corn Belt." *Ecological Economics* 134 (2017): 227-37.

Worku, Z., & Elmar, C. (2016). Identifying Categorical Land Use Transition and Land Degradation in Northwestern Drylands of Ethiopia. *Remote Sensing, 8*(5), 1-20, DOI:10.3390/rs8050408

Abbreviations: FD, food desert; ABM, agent-based model; ASD, agricultural statistical districts; BF, berries & non-tree fruits; CDL, cropland data layer; CT, census tract; DC, double season row crops; F, fallow; FL, farmland; GIS, geographical information system; H, hallow; LUT, land-use transition; NASS, national agricultural statistical services; NC, north carolina; NCCPI, national commodity crop productivity index; NFD, non-food desert; NRCS, national resources conservation services; NSF, national science foundation; SC, single season row crops; SPL, soil productivity layer; USDA, united states department of agriculture; VP, vegetables & produce.

ASSESSMENT OF THE PRECISION OF A SMART-PHONE POLE PHOTOGRAMMETRY FOR A SECOND-ORDER CLIFF SURFACE DEFORMATION STUDIES

M.A. Eboigbe[1], D. B. Kidner[2]

[1] Ph.D. Student, University of South Wales, Faculty of Engineering and Computing, UK. mitchell.eboigbe@southwalees.ac.uk
[2] Reader, University of South Wales, Faculty of Engineering and Computing, UK david.kidner@southwales.ac.uk.

KEY WORDS: Pole Photogrammetry, Coastal Cliff, Deformation Study

ABSTRACT:

Coastal cliff is almost a vertical elongated structure with a wave-cut notch and a landslip. Cliffs are geological formations with an almost unpredictable and unstoppable detachment between constitutes formations. Due to health, safety, environmental, and military restrictions, there are more regulations and restrictions on the use of drones. There are also the issues of portability and high cost for the purchase of hybrid drones and Terrestrial Laser Scanners (TLS). These negate the regular monitoring of the coastal cliff. This research develops a rapid, low-cost, and precise digital photogrammetry methodology for the continuous monitoring of the cliff by using the pole as the platform and a mobile phone as a sensor. The most practical vertical camera angle, image overlaps, survey distance to the cliff, and realistic time range for surveys are all determined from the basic surveying principles. Precise geometrically related point clouds generated are with or without the Global Navigation Satellite Systems (GNSS). The standard deviation for "alignment and surface deviation" at every point on each point cloud is ± 0.05m in the Northing and ± 0.12m on the Easting's for the self-calibrated digital camera and without the use of GNSS control points. With the GNSS controls, the maximum deviation in the XYZ coordinates is ±5 cm. Change analysis performed identifies areas of cut, fill, and the segment of threats in all point clouds. The photogrammetric technique developed is very cheap, simple, and reliable with minimum labor. The results obtained indicate the applicability of this methodology for second-order cliff Deformation study.

1. INTRODUCTION

With drones, pre-programmed flight routes are automatically set at the best angle that captures images on a regular grid pattern and flight speed (Varela et al, 2019; Colomina and Molina, 2014; Perez-Alberti and Trenhaile, 2014). To survey the cliff surface on a manual flight mission with the drone requires great experience to obtain an optimal flight mission and generate orthometric and digital models required for mapping purposes (Mancini et al, 2017). Digital images acquired arbitrarily from the ground stations to survey the cliff would be difficult to generate a complete overlapping stereo oriented model for consistent and precise 3D models and 3D maps except as a rare 'accidental success'.

The use of poles and masks as platforms would require ground-based camera locations. The distance between these ground-based stations and the cliff surface (L) is as the height (H) above the ground in aerial surveys. Like the traditional aerial surveys, camera-based stations on the ground would require very careful and systematic planning to achieve the best Ground Sampling Distance (GSD) (Huang et al, 2015), sensor stability (Roncella et al, 2014), overlap and reduce the effects of random errors and accumulation of systematic errors (Aber and Babb, 2018; James and Robson, 2014). The length between the baseline and cliff (L), the linear dimension of the pixel (px), and the camera lens' focal length (f) will determine GSD accuracies as camera pixels will be mapped to the cliff surface

Geological formations of limestone, sandstone, chalk, rock shelters, and granite characterize the cliff surface (Terefenko et al, 2018). Several crests of waves on the cliff surface depend on the formation of the sections on the cliff and such irregular spectral frequencies cause both low and high contrast across the electromagnetic spectrum (Young et al, 2016). Digital photogrammetry is more challenging when using low-quality complementary metal-oxide-semiconductor CMOS or the Charge-coupled devices CCD image sensors (Dai et al, 2014) as with this research. The best pattern of forward and side overlap and the distance between the camera's stations to the cliff surface is; therefore, determine to generate the best spatial and spectral resolutions for deformation analysis.

Due to the height of the coastal cliff and considering the very short distance between the camera location and the cliff surface, there must be careful considerations on the angle of the camera, camera calibration, stabilization, and sensor orientation to overcome yaw variation, rolls variation, altitude variation, perspective variation, barrel, and pincushion. The accuracies on the computation of the exterior orientation parameters by the BBA using the Integrated Sensor Orientation (ISO) is usually affected by both the satellite geometry and the lock on the available GNSS constellation (Rau et al, 2011). GNSS radio signals delays are mainly by the free electrons in the ionosphere and by other atmospheric conditions in the troposphere (Zhang et al, 2019). The Earth rotation induces charged particles into the magnetic field line in the ionosphere and capable of distorting the propagation of radio waves signal (Constable, 2016). Ionospheric propagation delays are of higher magnitude than the tropospheric delays and largely degrade the accuracies in single-frequency GNSS receivers' especially inbuilt GNSS sensors on smartphones and other inexpensive digital cameras (Karaim et al, 2018). Considerations are in place also to obtain the best time-lapse for satellite geometry when mapping the cliff surface using the pole photogrammetry due to drift in the GPS/INS low-quality digital camera/GPS sensor (Masiero et al, 2017).
The coastal cliff undergoes regular physical and chemical weathering processes (Earlie et al, 2017 and Sciarra et al, 2014)

as well as seawater crashing against the surface (Earlie et al, 2017). Cliff surface weathering causes an almost unpredictable and unstoppable detachment between constitutes formations and eventual collapse (Singh et al, 2016). The coastal cliffs are therefore asymmetrical with complex surface discontinuities (Letortu et al, 2018; Westoby et al, 2018; Ruzic et al, 2015; Galea et al, 2014; Somma et al, 2015 and Sciarra et al, 2014). It is difficult therefore to establish temporary/permanent control points on the cliff surface. The establishment of GNSS points close to the cliff at either the top or toe in some coastal area is practically difficult due to such high-volume sea wave and cliff surface erosion (Earlie et al, 2017). This is the situation in Penarth (study location) very peculiar with the Bristol Channel and the regular cliff face collapse. GNSS equipment is expensive to buy/ borrow and may not be available in most situations. It is imperative to develop a cost-effective and precise methodology using the Structure from Motion SFM derived point cloud for coastal cliff geotechnical and structural change detection analysis. For existing close-range digital photogrammetry techniques, consistent and accurate time-series orthomosaics and digital models for environmental monitoring are feasible on the availability of GNSS points (Ruzgiene et al, 2015 and Turner et al, 2015). The method developed is a simple and straightforward photogrammetry technique for consistent and precise evaluation of the cliff surface with or without the GNSS controls.

2. METHODOLOGY

2.1 Determine an Appropriate Vertical Camera Angle to the Cliff

The basic transformational matrixes of each grid cell relate every pixel location in the image relative to the cliff surface elevation. The digital camera self-calibration solution includes the direct computation for the interior and exterior parameters, which allows the transformation from 3D coordinate (X, Y, Z) of the ground point to a 2D coordinate (u, v) of the image in pixels. The basic perspective projection centre of the perspective lens cameras relates the scene properties as pixels where the origin is the camera centre defined by its position and orientation.

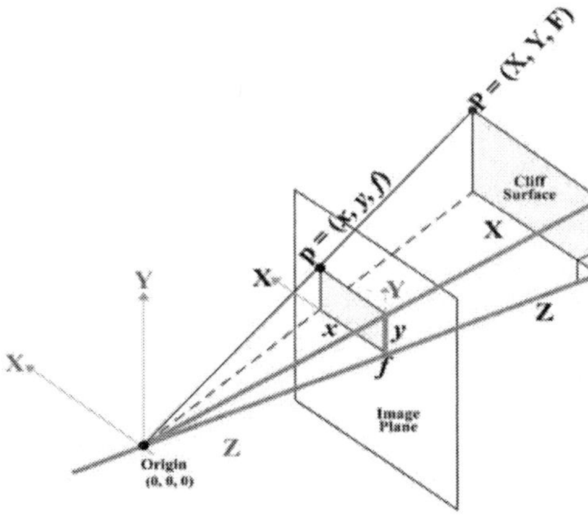

Figure 1. Perspective Projection

The origin and geometrical sequence between each grid cell determine the transformation matrix of orientation parameters that form the image of the object by relating the scene properties as pixels (Liu et al, 2016). The origin is the camera centre defined by its position and orientation (Xu et al, 2017).

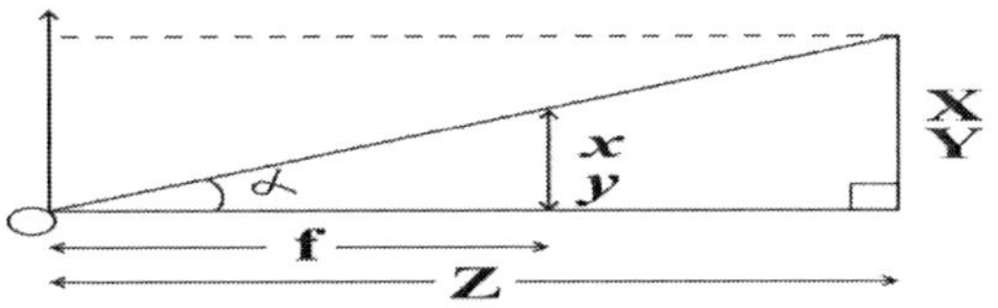

Figure 2. Central Projection

$$x = f\frac{X}{Z} \text{ and } y = f\frac{Y}{Z} \qquad (1)$$

Digital photos acquired from off-nadir angle geometry will generate denser tie-points and create a better model for the cliff surface compared with photos acquired from the nadir positions (Nesbit and Hugenholtz, 2019; Juad et al, 2016; Mancini et al, 2017). The reduction of systematic errors is from the reduction of gaps between adjoining digital images on the cliff surface geometry that is vertically irregular (Nesbit and Hugenholtz, 2019). This section experiment on an appropriate vertical camera angle that would be suitable to;

a) High overlapping images that can generate a dense point cloud
b) Obtain pixel texture accuracies for cliff surfaces
c) Survey the entire height of the cliff with geometric accuracies

This practical solution is reliant on the cliff surface characteristics, the resolution of the digital camera, and the pole as a platform using the perpendicular orientation geometry as shown below.

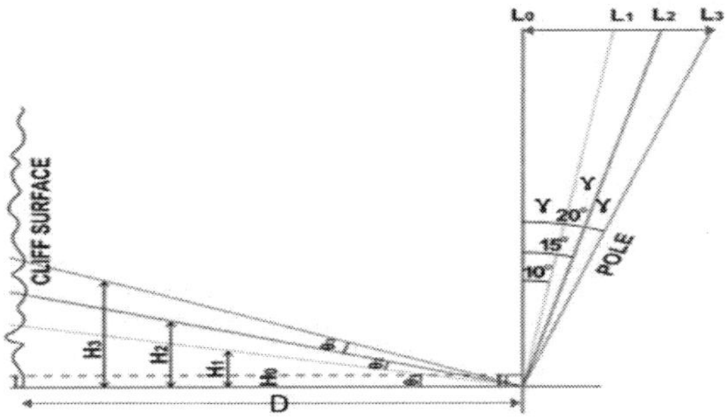

Figure 3. Viewing and Orientation Geometry to the Cliff

Θ = Observing angle
ϒ= Vertical camera angle
H = Height of the section on the cliff
D = Ground distance
Pϒ = Pixel size perpendicular to the vertical camera angle
PΘ = Angular segment on the cliff surface. $\tan \Theta = \frac{H}{D} =$ **(2)**

At L, the height of the cliff will be H while Pϒ is smaller provided every other photogrammetric constraint such as clouds, haze, and poor image overlap are minimal. This research experiments on the perpendicular orientation geometry using four different surveys at four different vertical cameras angle to the cliff section measuring 27 meters in length. The outcome is figure 4 below.

https://doi.org/10.5194/isprs-archives-XLIV-M-2-2020-15-2020 | © Authors 2020. CC BY 4.0 License.

Figure 4. Dense Point Cloud obtained from Four Vertical Camera Angle

2.2 Determining the Best Distance between the Cliff Surface and the Base Line

In close-range photogrammetry, the digital camera could be as close as a few centimeters to the object being monitored which favorably improves on the image resolution (Westoby et al, 2015). However, when monitoring an unstable cliff like the Penarth, 1-2 meters could be dangerous and unfavorable to the field of view (FOV) with constraints on the orientation angle of the digital camera. The best practice would be a trade-off between the distance to the cliff surface, the FOV, and the cliff height. The camera distance to the cliff surface determines the GSD. Shorter distances between the camera and the cliff will turn out into a smaller GSD provided every other photogrammetric constraint is at minimal while larger distances could amount to greater GSD especially with the use of low/medium resolution digital cameras. Smaller GSD will have a higher spatial resolution. The GSD should be at least half of the smallest object on the ground for measurement (Tziavou et al, 2018).

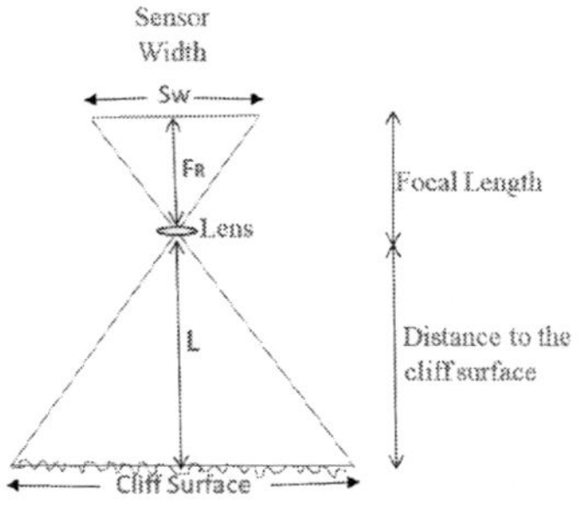

Figure 5 Sensor Width Relative to the FOV

$$GSD = \frac{L * px}{f}$$

$$L = \frac{GSD * F}{PX} \quad (3)$$

Most times the pixels, when projected onto the ground, will not be squared perfectly. The GSD would then be computed for in terms of sensor width and sensor height.

GSD = Ground Sampling Distance
L = the length between the baseline and cliff
f= the focal length of the digital camera
px = the pixel linear dimension

To determine approximately the GSD For a distance of 3 meters between the baseline and the cliff, using the Samsung (SM-G850F Camera);
The focal length of the Samsung SM-G850F camera = 1.2 mm
Pixel height = 1920
Pixel width = 1080
Approximate sensor width = 9.61923 mm
Approximate sensor height = 5.410082 mm

$$\text{Width} = \frac{3 * 5.410821}{1.2 * 1080} = \mathbf{1.25 cm/px} \qquad \text{Height} = \frac{3 * 9.619238}{1.2 * 1920} = \mathbf{1.25 cm/px.}$$

To derive a baseline that is photogrammetric stable between the camera position and the cliff, the cliff was surveyed three different times at three different baselines. The section of the cliff monitored is 27 meters in length. The Samsung Alpha phone SM-G850F Front Camera of 2.1 megapixels was on an 11 meters pole for these surveys. The outcome of each survey is the table I below

L (m)	No. of Photos	Average GSD (cm)	2D Key points extracted for BBA	3D Key points for BBA	Mean projection errors in pixel
3	169	1.68	3782815	1303566	0.254
5	169	0.73	3174090	786342	0.268
10	169	2.05	3958836	838557	0.309

Table 1. The outcome of three Different Surveys to determine a Practical Base Line to the Cliff

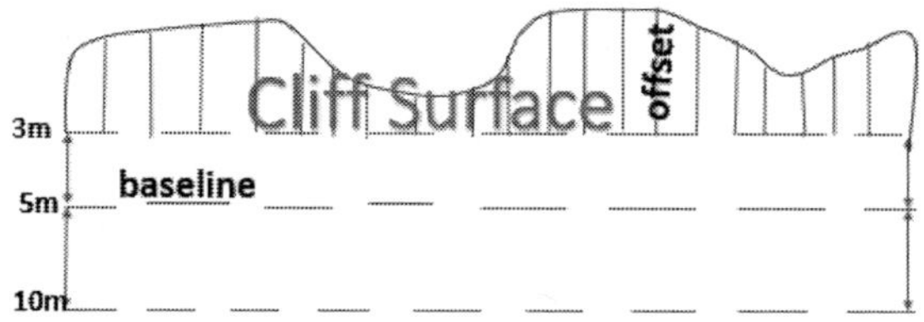

Figure 6. The Three Different Baseline Distances from the Cliff

2.3 Determine a Practicable forward and side overlap

The terrain topography and purpose of the photogrammetric project determine the choice and percentage of overlap (Pepe et al, 2018). Recommendations, however, are for a regular grid pattern (Carter et al, 2019) of 60% side overlap and 80% forward overlap (Cwiakala et al, 2018; Turner et al, 2015; Colomina and Molina, 2014). To generate tie-points with maximum density, every point in the object space should be in a minimum of 4-5 images (Ruzgiene et al, 2015).

https://doi.org/10.5194/isprs-archives-XLIV-M-2-2020-15-2020 | © Authors 2020. CC BY 4.0 License.

Figure 7. Degree of Overlap from Marked Points on the Ground

Poles stations were at every 1.65m marked on the ground with a white paper as shown on the Left-Hand Side LHS of fig. 4.9. The painters' pole used is 11m long and segmented at every 1m. The average percentage of overlap is 55% horizontal and 58% vertical. Figure 5 below illustrates the number of overlapping images.

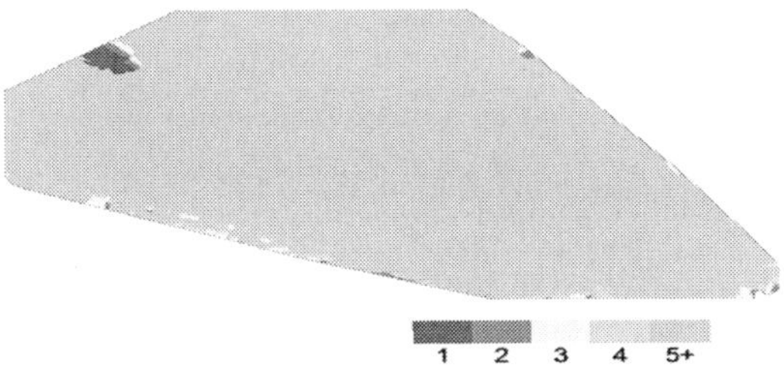

Figure 8. Number of Overlapping Images

2.4 Determine an Appropriate Time-Lapse for Surveying the Cliff with the Pole Using the Integrated Sensor Orientation (ISO)

In the ISO, the computation of the Bundle Block Adjustment BBA is by the use of the inbuilt GNSS sensors and the compact Micro-Electro-Mechanical Systems Inertial Measurement Unit (MEMS IMU) without GNSS controls and pre-defined tie points (Pepe et al, 2018 and Benassi et al, 2017). Although the ISO is straight to use, the cheap miniature sensors with cheap inbuilt GPS are downgraded easily especially in very dynamic environments (Benassi et al, 2017). The accelerometer and gyros of the MEMS-IMU are also flexible and can easily drift during use due to their microstructure compartments. The tropospheric signal delays characterized by pressure, temperature, and water vapour are relative to geographical elevation (Yu et al, 2018 and Zhu et al, 2017) with water vapour predominant in the coastal areas. The accuracies of close-range photogrammetric surveys will improve on favourable atmospheric conditions (Pepe et al, 2018) likely on lesser time durations.

Inexpensive GNSS sensors would be difficult to generate consistent and reliable dense point cloud due to inherent anomalies in the Scale Invariant Feature Transformations (SIFT) (Qu et al, 2018) which detects and extract the local features in the corresponding digital images. It would be difficult then to minimize errors from transformation and misalignments which ideally are reduced by the GNSS controls and tie points as with the Integrated Sensor Orientation (ISO) and GNSS-supported Aerial Triangulation (GNSS-AT) in digital photogrammetry (Benassi et al, 2017). To balance off systematic and random errors from the ionospheric and tropospheric delays, drifting, and downgrading of the inbuilt GPS, this research relies on the systematic field procedure(s) developed in the foregoing sections. The aim is to obtain the maximum number of images that would be calibrated for SIFT. The cliff surface was surveyed at different time-range to establish the level-of-tolerance of the inexpensive Samsung SM-G850F MEMS-IMU/inbuilt GNSS sensors on favourable atmospheric and illumination (shadow cast) conditions.

	Time Range Minutes	No. of images	% of calibrated images	Using the New field Method	Point cloud
a	00:30:03	86	100	yes	Single block mosaic
b	00:59:28	187	100	yes	Single block mosaic
c	01: 28:20	198	100	yes	Single block mosaic
d	02:16:31	490	98	80%	Multiple block mosaic
e	02:52:50	377	100	yes	Single block mosaic
f	03:05:36	517	99	yes	Single block mosaic

Table 2. Time Range to Determine the Level-of Tolerance of the Miniature Sensors

While surveys executed at lesser times will easily overcome systematic and random errors, good overlap, and a regular survey grid pattern can compensate for errors. The pattern for capturing images in **d** on table 2.was a horizontal survey grid with more images taken after the main survey to verify the effect of drifting. The dense point cloud from survey **d** in table 2. is downgraded into multiple block mosaics as shown below.

Figure 8. Multiple Block Point Cloud Showing the Effect of Drifting on the Inexpensive GNSS Sensors

3. EVALUATING THE GEOMETRIC PRECISIONS OF THE POLE SURVEY FOR CLIFF SURFACE DEFORMATION STUDIES

The different point clouds generated should be geometrically precise for evaluative purposes (Kromer et al, 2015). Comparisons were on four different point clouds obtained from four weekly surveys to determine the consistencies and reliability of the new pole mapping field sequence.

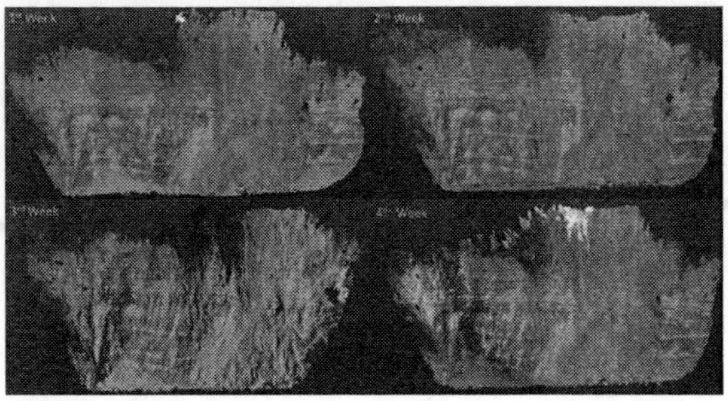

Figure 9. Dense Point Clouds from four Weekly Surveys

	1st Survey	2nd Survey	3rd Survey	4th Survey
Date of survey	02-10-18	11-10-18	22-10-18	30-10-19
N0. of images	409	461	377	496
Area covered	0.0033km^2	0.0029km^2	0.0018km^2	0.0035km^2
Average density (m^3)	12646	21118.3	25141.3	14089.7

Table 3. Procedure for the four Weekly Surveys

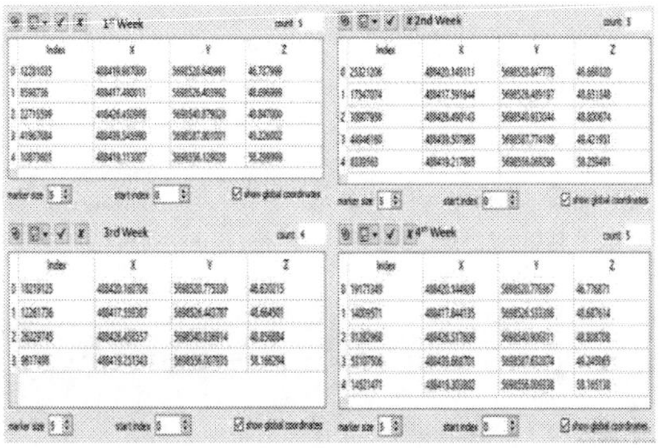

Figure 10. Geographical Coordinates of all four Weekly Surveys in CloudCompare

	Wk 1 –WK 2		Wk 1 –WK 3		Wk 1 –WK 4	
	ΔN	ΔE	ΔN	ΔE	ΔN	ΔE
Pt.1	-0.158111	-0.206787	-0.173706	-0.134339	-0.157928	-0.135376
Pt.2	-0.111633	-0.085205	-0.079376	-0.039795	-0.164124	-0.12939
Pt. 3	-0.039154	-0.054016	-0.007568	0.042114	-0.06662	-0.02728
Pt. 4	0,038025	0.126892	-0.138336	0.121093	-0.122711	0.20813
Pt. 5	-0.104858	0.06073			-0.190795	0.11969

Table 4. Accuracies of all four Weekly Surveys

The standard deviation for "alignment and surface deviation" (Westoby et al, 2018) at every point on each point cloud is ± 0.05m in the Northing and ± 0.12m on the Easting's for the self-calibrated digital camera and without the use of GNSS control points. This shows that each of the point cloud models geometrically fits into the other for a second-order cliff monitoring deformation study (Westoby et al, 2018 and Ruzic et al, 2015).

3.1 **Verifying the Correctness of the CloudCompare Global Coordinates**

To verify the correctness of the CloudCompare coordinates, the DSM of one of the Pole Survey Point Cloud (PSPC) was imported into ArcMap and the coordinates of the control points in CloudCompare as in Figure 10 is found to be the same as in ArcMap.

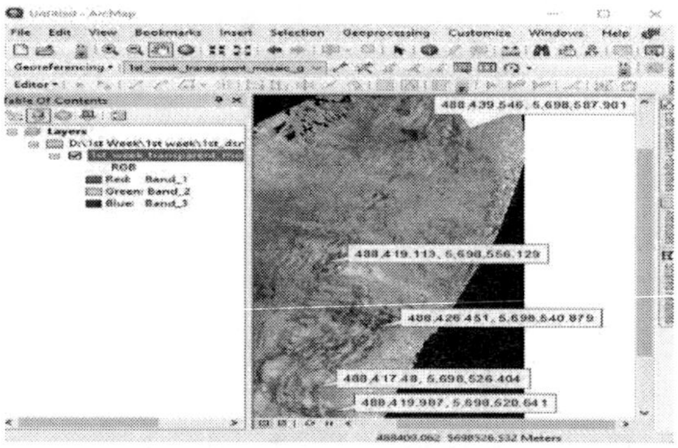

Figure 11. Verification of the CloudCompare Global Coordinates in ArcMap

The geometric accuracy defines the comparative evaluation with datasets obtained from other reliable photogrammetric techniques while the absolute accuracy is the difference of the x, y, and z coordinates of every point on the point cloud to their true geographical position on ground (Harwin and Lucieer, 2012). The 3D reconstruction process in the bundle adjustment may distort the original shape of the cliff surface alignment.

The TLS and the Pole survey were both used to survey the same stretch of the cliff surface on the 13th of March 2020. This is to be able to determine the global correctness of the digital image EXIF and structure from motion computation for bundle adjustment that optimizes the 3D location between the tie-points and the camera internal parameters (Turner et al, 2012).

The TLS survey was using the Topcon Laser scanner GLS-2000M series. The 8.9-degree telephoto camera provided the complete stereoscopic coverage of the cliff surface at 6.3 mm at 10 m resolution. Two different prism scan locations provided the precise and accurate modelling of the cliff section monitored. The fixed pole at (window scanning) enhances accuracies although it takes longer to scan but reduces the registration of the point cloud in the Topcon ScanMaster software. The height of the pole for this survey was 2 m and the prisms are 125 mm. Georeferencing was by the GEOMAX Zenith 35 series on Real-Time Kinematic (RTK) mode.

This contribution has been peer-reviewed.
https://doi.org/10.5194/isprs-archives-XLIV-M-2-2020-15-2020 | © Authors 2020. CC BY 4.0 License.

3.2 Control Points for Georeferencing the Pole Survey

Four markers at the bottom of the cliff with other visible points on both surveys in the middle of the inaccessible clifftop were used as manual tie-points for geolocating the pole survey in Pix4D on OSGB 1936 coordinate system as shown in figure 12 below.

Figure 12. Control Points for Geolocating the Pole Survey

Figure 13. Pole Survey Point Cloud

The X, Y, and Z coordinates of some points at the top, sides, and at the bottom of the pole and TLS point clouds, read off, and compared in CloudCompare as shown in figure 15 and table 5 below.

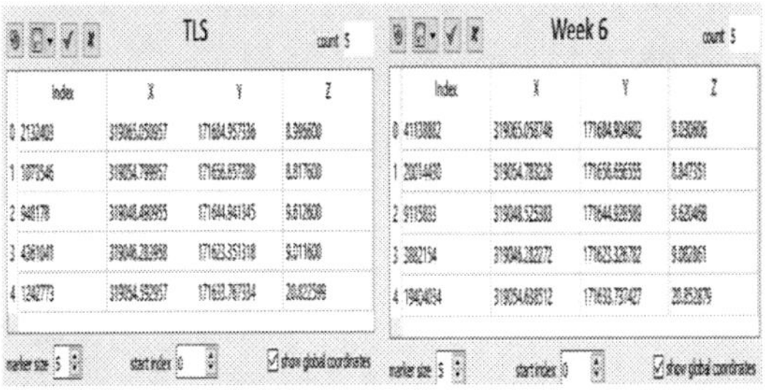

Figure 15. Example of Coordinates of TLS and Pole Survey in CloudCompare

Table 5. Differences in the Coordinates of some Points in the TLS and Pole Surveys

	TLS – Week 6		
	ΔN	ΔE	ΔZ
Point1	-0.008	0.053	-0.035
Point 2	0.017	0.001	-0.030
Point 3	-0.034	0.013	-0.008
Point 4	0.002	0.024	-0.071
Point 5	-0.045	0.030	-0.030

The maximum deviation in the XYZ coordinates of the PSPC at every point as compared with the TLSPC is 5 cm.

The accuracies of the geometry of the PSPC and the TLSPC are subject to further scrutiny by creating their contours using the triangulation and the linear interpolation gridding method in the Surfer Golden Software (Sufer, 2018). The triangulation with linear interpolation uses the anisotropy to determine its contour (Bayazit, 2019). Anisotropy is dependent on the direction irrespective of the properties of the surface materials will generate an unbiased contour interpretation (Bayazit, 2019). The choice to use the multiple refractive indexes is due to the irregular surface nature of the cliff that has the capabilities for change in direction within the shortest possible period resulting from falls and collapse.

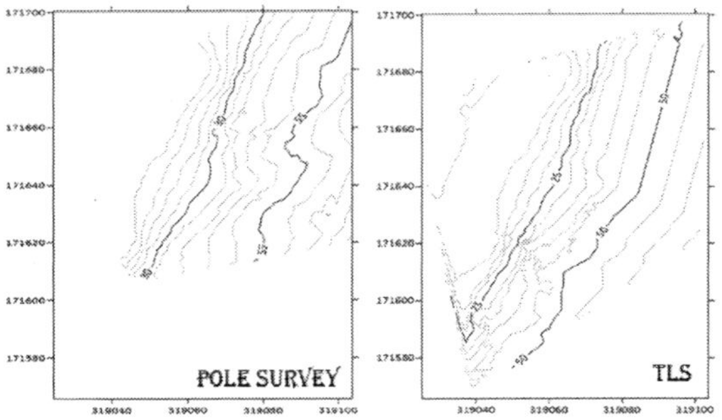

Figure 16. PSPC and TLSPC Contours generated in Suffer Golden Software

4. DISCUSSION OF RESULTS

4.1 Visual Comparison with Other Datasets

Figure 17. Pole Survey (on white) Superimposed on an Equivalength Drone Survey

Figure 18. Like-to-like Alignment (Pole Survey up and Drone Survey down)

https://doi.org/10.5194/isprs-archives-XLIV-M-2-2020-15-2020 |

Figure 19. 2018 Google Earth Imagery of the Study Area

Recent studies show that the M3C2 will be more effective for comparing irregular surfaces like the cliff (Lague et al, 2013). This is because the algorithm distance computation is not dependent on the spatial correlation and the dimensional orientation between the surfaces (Esposito et al, 2017).

In figure 20 below, the M3C2 in the open-source Cloudcompare is used to perform change analysis between four weekly surveys. The evaluation of the detectable surface changes is at a minimum of ±6mm. From week 1 to 2, the fuller Earth is obvious and completely erodes from week 4 to week 5. There are insignificant changes from week 2 to week 3 and complete cliff erosion at points a. b, c, d, e, f, g, h, I.

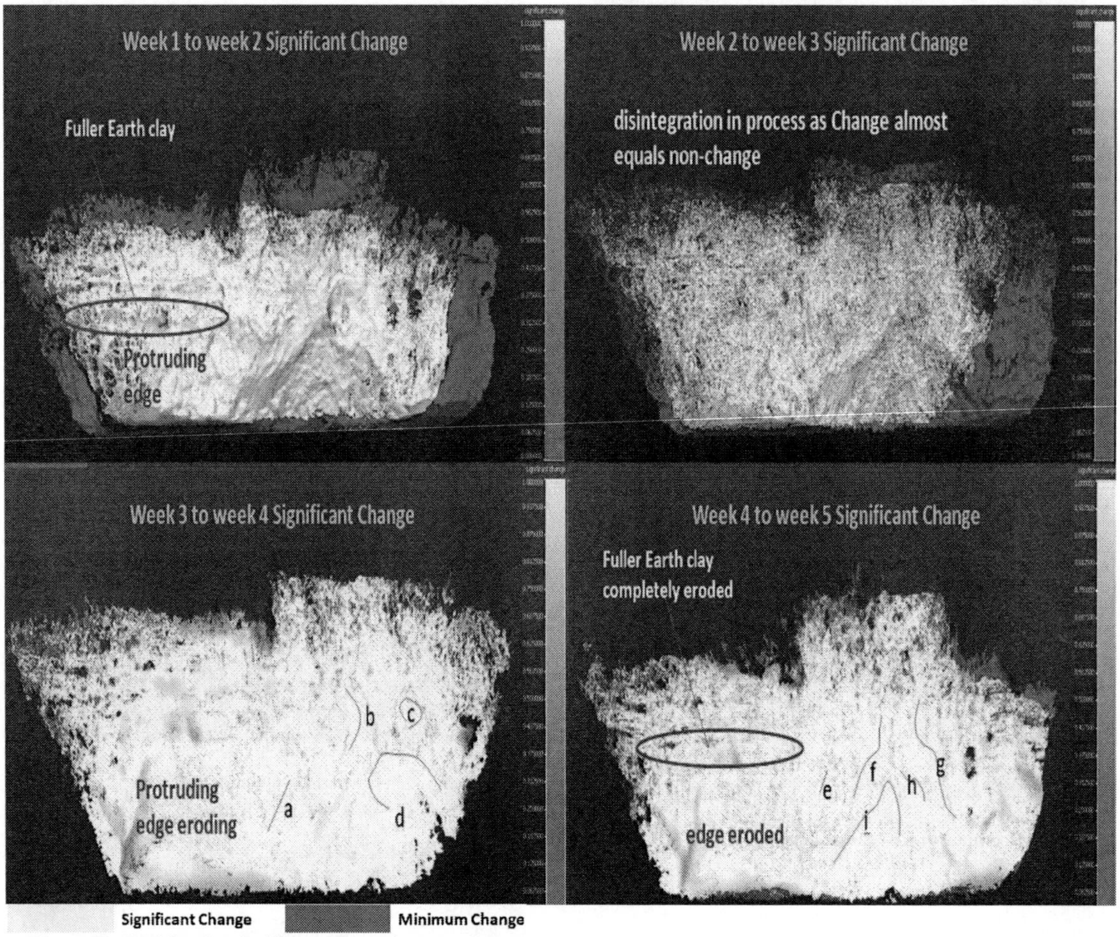

Figure 20. Explains the significant changes between four weekly surveys at 6mm detectable surface change.

5. CONCLUSION

The contribution to knowledge is a new pole digital photogrammetry methodology for coastal cliff monitoring.
This research utilizes the very basic principles of photogrammetry in developing a systematic flight plan and camera orientation to able to generate precise and accurate 3D reconstructed point cloud by terrestrial pole survey for coastal cliff monitoring. The methodology developed is simple, straightforward with a regular grid network for complete stereoscopic coverage. The verification of results obtained has been by third-party software and an alternative remote sensing survey.

This methodology generates precise geometrical related point clouds with or without the GNSS controls. For use in the photogrammetry community, the most appropriate vertical camera inclination and distance to the cliff surface has been practically demonstrated and theoretically documented.

https://doi.org/10.5194/isprs-archives-XLIV-M-2-2020-15-2020 | © Authors 2020. CC BY 4.0 License.

Depending on the area been mapped, surveys can be from 10 minutes to 2 hours. It is possible to attach the different digital cameras on the pole with the higher resolution cameras resulting in higher spectral accuracies. The limitation is to perform a survey in an almost equal atmospheric condition to avoid the drifting of the MEMS-IMU during the survey.
Surveys performed at a very close range can detect hidden sections/crack on the cliff, but care must be not to change the viewing angle of the digital camera during use. Surveys do not require the drone licence, but care must be not to drop the pole during use.

REFERENCES

Aber, J. W. and Babb, T. A. (2018) 'The Challenges of Processing Kite Aerial Photography Imagery with Modern Photogrammetry Techniques', *International Journal of Aviation, Aeronautics, and Aerospace* 5(2) pp. 1-30. *IJAAA* [Online]. Available at: https://commons.erau.edu/ijaaa/vol5/iss2/2/ (Accessed: 25 March 2019).

Bayazit, U. (2019) 'A Greedy Region Growing Algorithm for Anisotropic Stretch Adaptive Triangulation of Geometry Images', *Graphical Models* 106(101045) pp. 1524-0703. *ScienceDirect* [Online]. Available at: https://www.sciencedirect.com/science/article/pii/S152407031 9300360 (Accessed: 13 June 2020).

Benassi, F., Dall'Asta, E., Diotri, F., Forlani, G., Morra di Cella, U., Roncella, R. and Santise, M. (2017) 'Testing Accuracy and Repeatability of UAV Blocks Oriented with GNSS-Supported Aerial Triangulation', *Remote Sensing* 9(2) pp. 2-23. *MDPI* [Online]. Available at: https://www.mdpi.com/2072-4292/9/2/172 (Accessed: 11 December 2017).

Carter, N., Hashemian, A., Mckelvey, N. (2019) 'An Optimization of Small Unmanned Aerial System (sUAS) Image Based Scanning Techniques for Mapping Accident Sites', *SAE Technical Paper 2019-01-0427.SAE Mobilus* [Online]. Available at: https://saemobilus.sae.org/content/2019-01-0427/ (Accessed: 11 September 2019).

Colomina, I. and Molina, P. (2014) 'Unmanned aerial systems for photogrammetry and remote sensing: A review', *ISPRS Journal of Photogrammetry and Remote Sensing* 92, pp. 79-97. *Science Direct* [Online]. Available at: http://www.sciencedirect.com/science/article/pii/S0924271614 000501(Accessed: 25 May 2016)

Constable, C. (2016) 'Earth's Electromagnetic Environment', *Surveys in Geophysics* 37(1) pp. 27-45. *Springer* [Online]. Available at: https://link.springer.com/article/10.1007/s10712-015-9351-1 (Accessed: 03 October 2019).

Cwiakala, P., Kocierz, R., Puniach, E., Nedzka, M., Mamczarz, K., Niewiem, W. and Wiacek, P. (2018) 'Assessment of the Possibility of Using Unmanned Aerial Vehicles (UAVs) for the Documentation of Hiking Trails in Alpine Areas', *Sensors* 18(1) pp. 2-28. *NCBI* [Online]. Available at: https://www.ncbi.nlm.nih.gov/pmc/articles/PMC5795845/ (Accessed: 02 July 2019).

Dai, F., Feng, Y. and Hough, R. (2014) 'Photogrammetric Error Sources and Impacts on Modelling and Surveying in Construction Engineering Applications', *Visualization in Engineering* 2(2) pp. 2-14. *Springer* [Online]. Available at: https://link.springer.com/content/pdf/10.1186%2F2213-7459-2-2.pdf (Accessed: 17 December 2018). 04017008-1 - 04017008-21.

Earlie, C., Masselink, G. and Russel, P. (2017) 'The Role of Beach Morphology on Coastal Cliff Erosion Under Extreme Waves', *Earth Surface Processes and Landforms* 43(6) pp. 1213-1228. *Wiley Library* [Online]. Available at: https://onlinelibrary.wiley.com/doi/full/10.1002/esp.4308 (Accessed: 15 February 2018).

Esposito, G., Salvini, R., Matano, F., Sacchi, M., Danzi, M., Somma, R. and Troise, C. (2017) 'Multitemporal Monitoring of a Coastal Landslide through SFM-Derieved Point Cloud Comparison', *The Photogrammetric Record* 32(160) pp. 459-479. *Ingenta Connect* [Pnline]. Available at: https://www.ingentaconnect.com/content/asprs/pers/2010/00000 076/00000010/art00001;jsessionid=2b0icttstd7nn.x-ic-live-02# (Accessed 11 July 2020).

Galea, P., Amico, S. D. and Farrugia, D. (2014) 'Dynamic Characteristics of an Active Coastal Spreading Area Using Ambient Noise Measurements – Anchor Bay, Malta', *Geophysical Journal International* 199(2) pp. 1166-1175. *IEEE Xplore* [Online]. Available at: https://ieeexplore.ieee.org/document/8205841 (Accessed: 17 November 2017).

Harwin, S. and Lucieer, A. (2012) 'Assessing the Accuracy of Georeferenced Point Clouds Produced Via Multi-View Stereopsis from Unmanned Aerial Vehicle (UAV) Imagery', *Remote Sensing* 4(6) pp. 1573-1599. *MDPI* [Online]. Available at: http://www.mdpi.com/2072-4292/4/6/1573 (Accessed: 19 October 2016).

Huang, S., Zhang, Z., Ke, T., Tang, M. and Xu, X. (2015) 'Scanning Photogrammetry for Measuring Large Targets in Close Range', *Remote Sensing* 7(8) pp. 10042-10077. *MDPI* [Online]. Available at: https://www.mdpi.com/2072-4292/7/8/10042/htm (Accessed: 08 February 2019).

James, M. R. and Robson, S. (2014) 'Mitigating Systematic Error in Topographic Models Derived from UAV and Ground-Based Image Networks', *Earth Surface Processes and Landforms* 39(10) pp. 1413-1420. *Wiley Library* [Online]. Available at: https://onlinelibrary.wiley.com/doi/full/10.1002/esp.3609 (Accessed 07 February 2019).

Juad, M., Passot, S., Bivic, R. L., Delacourt, C., Grandjean, P. and Dantec, N. L. (2016) 'Assessing the Accuracy of High-Resolution Digital Surface Models Computed by PhotoSCan and MicMac in Sub-Optimal Survey Conditions', *Remote Sensing* 8(6), 465; pp. 2-18. *MDPI* [Online]. Available at: http://www.mdpi.com/2072-4292/8/6/465 (Accessed: 26 May 2018).

Karaim, M., Elsheikh, M. and Noureldin, A. (2018) '*GPS* GNSS Error Sources' inRustamov, R. B. and Hashimov, A. M. (eds), *Multifunctional Operation and Application of GPS,* IntechOpen, pp. 69-85. Available at: https://www.intechopen.com/books/multifunctional-operation-

and-application-of-gps/gnss-error-sources (Accessed: 03 October 2019).

Kromer, R. A., Abellan, A., Hutchinson, D. J., Lato, M., Edwards, T. and Jaboyedoff, M. (2015) A 4D Filtering and Calibration Techniques for Small-Scale Point Cloud Change Detection with a Terrestrial Laser Scanner, *Remote Sensing* 7(2) pp. 13029-13052. *MDPI* [Online]. Available at: https://www.mdpi.com/2072-4292/7/10/13029 (Accessed: 06July 2020).

Lague, D., Brodu, N. and Leroux, J. (2013) 'Accurate 3D Comparison of Complex Topography with Terrestrial Laser Scanner: Applications to the Rangitikei Canyon (N-Z)', *ISPRS Journal of Photogrammetry and Remote Sensing* 82 pp. 10-26. *Science Direct* [Online]. Available at: https://www.sciencedirect.com/science/article/abs/pii/S0924271613001184 (Accessed: 24 November 2017).

Letortu, P., Juad, M., Grandjean, P., Ammann, J., Costa, S., Maquaire, O., Davidson, R., Dantec, N. L. and Delacourt, C. (2018) 'Examining High-Resolution Survey Methods for Monitoring Cliff Erosion at an Operational Scale', *Behavioural Science and Public Health Titles* 55(4) pp. 457-476. *Hal. Archives* [Online]. Available at: https://hal.archives-ouvertes.fr/hal-01647588/document (Accessed: 31 August 2018).

Liu, S., Zhu, W., Zhang, C. and Sun, W. (2016) '3D Reconstruction of Indoor Scenes Using RGB-D Monocular Vision', *International Conference on Robots and Intelligent System (ICRIS),* Zhangjiajie, China, 27-28 August. *IEEE Xplore* [Online]. Available at: https://ieeexplore.ieee.org/abstract/document/7757064/authors#authors (Accessed: 10 September 2019).

Mancini, F., Castagnetti, C., Rossi, P., Dubbini, M., Fazio, N, L., Perrotti, M. and Lollino, P. (2017) 'An Integrated Procedure to Assess the Stability of Coastal Rocky Cliffs: From UAV Close-Range Photogrammetry to Geomechanical Finite Element Modelling', *Remote Sensing* (9)12 pp. 2-21. *MDPI* [Online]. Available at: https://www.mdpi.com/2072-4292/9/12/1235 (Accessed: 30 January 2019).

Masiero, A., Fissore, F. and Vettore, A. (2017) 'A Low Cost UWB Based Solutions for Direct Georeferencing UAV Photogrammetry', *Remote Sensing* 9(5) pp. 2-21. *MDPI* [Online]. Available at: https://www.mdpi.com/2072-4292/9/5/414 (Accessed: 27 March 2019).

Nesbit, P. R. and Hugenholtz, C. H. (2019) 'Enhancing UAV-SFM 3D Model Accuracy in High-Relief Landscapes by Incorporating Oblique Images', *Remote Sensing* 11(3) pp. 2-24. *MDPI*[Online]. Available at: https://www.mdpi.com/2072-4292/11/3/239/htm (Accessed: 17th June 2019).

Perez-Alberti, A. and Trenhaile, A. S. (2014) 'An Initial Evaluation of Drone-Based Monitoring of Boulder Beaches in Galicia, North-Western Spain', *Earth Surface Processes and Landforms* 40(1) pp. 105-111. *Wiley Library* [Online]. Available at: https://onlinelibrary.wiley.com/doi/full/10.1002/esp.3654 (Accessed: 27 March 2019).

Pepe, M., Fregonese, L. and Scaioni, M. (2018) 'Planning Airborne Photogrammetry and Remote- Sensing Missions with Modern Platforms and Sensors', *European Journal of Remote Sensing* 51(1) pp. 412-436. *TandF* [Online]. Available at: https://www.tandfonline.com/doi/full/10.1080/22797254.2018.1444945 (Accessed: 04 October 2019).

Qu, Y., Huang, J. and Zhang, X. (2018) 'Rapid 3D Reconstruction for Image Sequence Acquired from UAV Camera', *Sensors* 18(1):225 pp. 2-20. *NCBI* [Online]. Available at: https://www.ncbi.nlm.nih.gov/pmc/articles/PMC5795716/ (Accessed: 11 May 2018).

Rau, J. Y., Habib, A. F., Kersting, A. P., Chiang, K. W., Bang, K. I., Tseng, Y. H. and Li, Y. H. (2011),'Direct Sensor Orientation of a Land-Based Mobile Mapping System', *Sensors* 11(7) pp. 7243-7261. *NCBI* [Online]. Available at: https://www.ncbi.nlm.nih.gov/pmc/articles/PMC3231683/ (Accessed: 23 January 2019).

Roncella, R., Forlani, G., Fornari, M. and Diotri, F. F. (2014) 'Landslide Monitoring by Fixed-Based Terrestrial Stereo-Photogrammetry', *ISPRS Technical Commission V Symposium*, Riva Del Garda, Italy, 23-25 June. ISPRS Annals of the Photogrammetry, Remote Sensing and Spatial InformationSciences, Volume 11-15. *ISPRS* [Online]. Available at: https://www.isprs-ann-photogramm-remote-sens-spatial-inf-sci.net/II-5/297/2014/isprsannals-II-5-297-2014.pdf (Accessed: 08 February 2019).

Ruzic, I., Benac, C., Marovic, I. and Ilic, S. (2015) 'A Stability Assessment of Coastal Cliffs Using Digital Imagery', *Acta GeotechnicaSlovenica* 12(2) pp. 25-35. *Digital Library of University of Maribor* [Online]. Available at: https://dk.um.si/IzpisGradiva.php?id=70850&lang=eng (Accessed: 24 January 2019).

Ruzgiene, B., Berteska, T., Gecyte, S., Jakubauskiene, E. and Aksamitauskas, C. V. (2015) 'The Surface Modelling Based on UAV Photogrammetry and Qualitative Estimation', *Measurement* 73 pp. 619-627. *ScienceDirect* [Online]. Available at: https://www.sciencedirect.com/science/article/pii/S0263224115002316 (Accessed: 24 January 2019).

Sciarra, N., Marchetti, D., Avanzi, G. D. and Calista, M. (2014) 'Rock Slope Analysis on the Complex Livorno Coastal Cliff (Tuscany, Italy)', *GFDQ* 37 pp. 113-130. *GFDQ* [Online]. Available at: http://gfdq.glaciologia.it/037_2_04_2014/ (Accessed: 24 January 2019).

Singh, P. K., Kainthola, A., Panthee, S. and Singh, T. N. (2016) 'Rockfall Analysis Along Transportation Corridors in High Hill Slopes', *Environmental Earth Sciences* 75(5) pp. 1-11. *Springer Link* [Online]. Available at: https://link.springer.com/article/10.1007/s12665-016-5489-5#citeas(Accessed: 06 September 2019).

Somma, R., Matano, F., Marino, E., Caputo, T., Esposito, G., Caccavale, M., Carlino, S., Luliano, S., Mazzola, S., Molisso, F., Sacchi, M., Troise, C. and Natale, G. D. (2015) 'Application of LaserScanning for Monitoring Coastal Cliff Instability in the Pozzuoli Bay, Coroglio Site, PosillipoHill,Naples', *Engineering Geology for Society and Territory* 5 pp. 687-690. *Springer Link* [Online]. Available at: https://link.springer.com/chapter/10.1007/978-3-319-09048-1_133 (Accessed: 31 October 2017).

This contribution has been peer-reviewed.
https://doi.org/10.5194/isprs-archives-XLIV-M-2-2020-15-2020 | © Authors 2020. CC BY 4.0 License.

Surfer (2014) *Contouring, Gridding and 3D Surface Mapping* Available at: https://www.goldensoftware.com/ (Accessed: 13 June 2020).

Terefenko, P., Wziatek, D. Z., Dalyot, S., Boski, T. and Lima-Filho, F. P. (2018) 'A High –Precision LiDAR-Based Methods for Surveying and Classifying Coastal Notches', *International Journal of Geo-Information* 7(8) pp. 2-16. *MDPI* [Online]. Available at: https://www.mdpi.com/2220-9964/7/8/295 (Accessed: 08 February 2019).

Turner, D., Lucieer, A. and Jong, S. M. D. (2015) 'Time Series Analysis of Landslide Dynamics Using an Unmanned Aerial Vehicle (UAV)', *Remote Sensing* 7(2) pp. 1736-1757. *MDPI* [Online]. Available at: https://www.mdpi.com/2072-4292/7/2/1736/htm (Accessed: 24 January 2019).

Turner, D., Lucieer, A. and Watson, C. (2012) 'An Automated Technique for Generating Mosaics from Ultra-High Resolution Unmanned Aerial Vehicle (UAV) Imagery, Based on Structure from Motion (SFM) Point Clouds', *Remote Sensing* 4(5) pp. 1392-1410. *MDPI* [Online]. Available at: https://www.mdpi.com/2072-4292/4/5/1392/htm (Accessed: 05 June 2020).

Tziavou, O., Pytharouli, S. and Souter, J. (2018) 'Unmanned Aerial Vehicle (UAV) Based Mapping in Engineering Geological Surveys: Considerations for Optimum Results', *Engineering Geology* 232() pp. 12-21. *ScienceDirect* [Online]. Available at: https://www.sciencedirect.com/science/article/abs/pii/S0013795217308670 (Accessed: 08 September 2019).

Varela, M. R., Patricio, A. R., Anderson, K., Broderick, A. C., DeBell, L., Hawkes, L. A., Tilley, D., Snape, R. T. E., Westoby, M. J. and Godley, B. J. (2019) 'Assessing Climate Change Associated Sea-Level Rise Impacts on Sea Turtle Nesting Beaches Using Drones, Photogrammetry and a Novel GPS System', *Global Change Biology* 25(2) pp. 753-762. *Wiley Library* [Online]. Available at: https://onlinelibrary.wiley.com/doi/abs/10.1111/gcb.14526 (Accessed 16 April 2019).

Westoby, M. J., Dunning, S. A., Woodward, J., Hein, A. S., Marrero, S. M., Winter, K. and Sugden, D. E. (2015) 'Sedimentological Characterization of Antarctic Moraines Using UAVs and Strcture-from-Motion Photogrammetry', *Journal of Glaciology* 61(230) pp. 1088-1102. *IGS* [Online]. Available at: https://www.igsoc.org/ (Accessed: 01 September 2019).

Westoby, M. J., Lim, M., Hogg, M., Pound, M. J., Dunlop, L. and Woodward, J. (2018) 'Cost-Effective Erosion Monitoring of Coastal Cliffs', *Coastal Engineering* 138 pp. 152-164. *ScienceDirect* [Online]. Available at: https://www.sciencedirect.com/science/article/pii/S0378383917303381 (Accessed: 25 August 2018).

Xu, G., Zheng, A., Li, X. and Su, J. (2017) 'Position and Orientation Measurement Adopting Camera Calibration by Projection Geometry of Plucker Matrices of Three-Dimensional Lines', *Scientific Reports* 7(44092) pp. 1-10. *Scientific Report* [Online]. Available at: https://www.nature.com/articles/srep44092 (Accessed: 10 September 2019).

Young, A. P., Guza, R. T., O'Reilly, W. C., Burvingt, O. and Flick, R. E. (2016) 'Observations of Coastal Cliff Base Waves, Sand Levels, and Cliff Top Shaking', *Earth Surface Processes and Landforms* 41 pp. 1564-1573. *Wiley* [Online]. Available at: https://onlinelibrary.wiley.com/doi/full/10.1002/esp.3928 (Accessed: 08 February 2019).

Yu, C., Zhenhong, L. and Penna, N. T. (2018) 'Interferometric Synthetic Aperture Radar Atmospheric Correction Using a GPS-Based Iterative Tropospheric Decomposition Model', *Remote Sensing of Environment*204 pp. 109-121. *Science Direct* [Online]. Available at: https://www.sciencedirect.com/science/article/pii/S0034425717305011 (Accessed: 04 April 2019).

Zhang, Z., Pan, S., Gao, C., Zhao, T. and Gao, W. (2019) 'Support Vector Machine for Regional Ionospheric Delay Modelling', *Sensors* 19(13) pp. 1-14. *MDPI* [Online]. Available at: https://www.mdpi.com/1424-8220/19/13/2947 (Accessed: 03 October 2019).

Zhu, B., Li, J. and Tang, W. (2017) 'Correcting Insar Topographically Correlated Tropospheric Delays Using a Power Law Model Based on ERA-Interim Reanalysis', *Remote Sensing* 9(8) pp. 1-23. *MDPI* [Online]. Available at: https://www.mdpi.com/2072-4292/9/8/765 (Accessed: 04 October 2019).

A METHOD TO GENERATE FLOOD MAPS IN 3D USING DEM and DEEP LEARNING

Asmamaw Gebrehiwot[1], Leila Hashemi-Beni[2*]

[1] Applied Science and Technology Ph.D. Program, Department of Built Environment, North Carolina A&T State University - aagebrehiwot@aggies.ncat.edu
[2*] Geomatics Program, Department of Built Environment, North Carolina A&T State University - lhashemibeni@ncat.edu

KEY WORDS: Floods, Remote Sensing, Convolutional Neural Networks, UAV, Structure from Motion, LiDAR

ABSTRACT:

High-resolution remote sensing imagery has been increasingly used for flood applications. Different methods have been proposed for flood extent mapping from creating water index to image classification from high-resolution data. Among these methods, deep learning methods have shown promising results for flood extent extraction; however, these two-dimensional (2D) image classification methods cannot directly provide water level measurements. This paper presents an integrated approach to extract the flood extent in three-dimensional (3D) from UAV data by integrating 2D deep learning-based flood map and 3D cloud point extracted from a Structure from Motion (SFM) method. We fine-tuned a pretrained Visual Geometry Group 16 (VGG-16) based fully convolutional model to create a 2D inundation map. The 2D classified map was overlaid on the SfM-based 3D point cloud to create a 3D flood map. The floodwater depth was estimated by subtracting a pre-flood Digital Elevation Model (DEM) from the SfM-based DEM. The results show that the proposed method is efficient in creating a 3D flood extent map to support emergency response and recovery activates during a flood event.

1. INTRODUCTION

Accurate information about the flood inundation extent and water depth is essential for relief activities. Flood risk maps can protect human lives and property damages by providing timely damage assessments to plan relief work efficiently. The use of remote sensing data to assess flood hazards in near real-time has been popular over the last few decades (Huang et al., 2018). Unmanned Aerial Vehicles (UAVs) have been rapidly developed as tools for inundation mapping tasks. UAVs survey offers a viable alternative to conventional systems because they deliver high-resolution data at a lower cost from complex urban landscapes, inaccessible areas, and hazardous environments (Feng et al., 2015, Boccardo et al., 2015).

Several methods have been proposed to create a 3D inundation map from the remote sensing imagery. Since the pioneering work of Miller et al. (1958), DEMs have been widely analyzed for flood extent mapping. DEMs can be produced from field survey data, satellite and aerial images, and Light Detection and Ranging (LiDAR) data. Generating DEM utilizing conventional field surveys and GPS needs more skilled manpower and is time-consuming and expensive to execute over a large area. LiDAR has been widely used to create DEMs since it can map large spatial areas by less manpower. LIDAR is a remote sensing technology that uses light in the form of a pulsed laser to measure ranges or variable distances to the Earth. LiDAR has a minimum human dependency, weather/ light independence, is a fast acquisition and processing approach for DEM creation. However, the LiDAR instrument cannot record elevation data from flooded areas since the lidar lights absorbed by water. SfMs solve multicamera viewing problem that generates high-density point cloud from high-resolution overlapping images (Caroti et al., 2015). SfM is the process of estimating the 3D structure of objects from the series of 2D images. The success of SfMs governed by the quality of the image resolution, degree of image overlaps, the relative motion of the camera with respect to the scene, and the presence of sharp edges (Watts et al, 2012, Remondino et al, 2014). While SfM photogrammetry is increasingly used for mapping in various applications, only a few researchers have used SfM for flood mapping, and its potential has not yet been fully explored by the remote sensing community. Meesuk, et al. (2012) investigated the efficiency of SfM approaches and achieved promising results in terms of accuracy for mapping flood extents. In their another work Meesuk, et al. (2015), they used the SfM technique to detect hidden urban features from a sequence of ground-view images that significantly improved the accuracy of flood simulation results. Hashemi-Beni et al. (2018) provided an overview of the opportunities and challenges of 3D flood extent mapping using UAV and SfM. The main challenges include (1) poor environmental and weather conditions (such as wind) during a flooding event ; (2) availability and visibility of ground control points (GCPs); (3) insufficient tie-points for image calibration due to the homogenous appearance of the water surface.; (4) needs for reclassification of noisy points cloud generated from SfM to determine the extent of the flood. Based on this context, we proposed an integrated method (SfM and Deep learning) to classify a 3D flood point cloud and generate a 3D flood map.

Deep neural networks such as Convolutional Neural Networks (CNNs) have been applied to many vision-related tasks for automatically extracting targets and demonstrated outstanding performance. CNNs applied to image classification (Ciregan et al., 2012), object detection (Girshick et al., 2014), scene labeling (Farabet et al., 2012), and object recognition (Taigman et al., 2014). The most notable advantage of CNNs is that they can construct high-level features from a raw dataset in an incremental manner. CNNs are widely used for remote sensing tasks because of their ability to successfully handle large training data sets, often yielding higher classification accuracy. Many research studies used CNN for automatically extracting a 2D flood extent (Gebrehiwot et al., 2019, Sarker et al., 2019). Gebrehiwot et al. (2019) modified VGG-based FCNs to extract the flooded extent from UAV images. The CNN-based model was trained using 100 manually labeled images and achieved more than 95% accuracy on extracting inundated areas. Sarker et al. (2019) proposed an FCN model for mapping flood extent in 2D from Landsat satellite images. A new satellite dataset from flooded areas across

Australia was used to evaluate the performance of the model. A comparison of their proposed model to the conventional support vector machines (SVM) classifier shows that the FCN is more accurate (76.7%) compared to SVM (45.27%) for the classification task. These deep learning-based studies are limited to extract 2D flooded areas and cannot provide water level measurements from 2D imagery. To the best of our knowledge, the potential of CNNs for 3D flood extent mapping has not been explored and verified.

This research attempts to fill this gap in the literature by exploring and verifying the potential of an integrated CNN and SfM approach for 3D inundation mapping. In this study, the 2D flooded areas will be automatically extracted using a deep learning method. The extracted flooded areas will be overlaid onto the 3D map generated by the SfM approach to determine the flood extent. The floodwater level will be estimated by subtracting the SfM-based DEM (water surface) with a pre-flood Lidar-based DEM.

The paper is organized as follows. In Section 2, the study areas and the data used for the research are described. Section 3 explains the approaches used to create a 3D inundation map. The results and discussion are presented in Section 4. Finally, we conclude by summarizing our results in Section 5.

2. STUDY AREAS AND DATA

Two flood-prone areas, Lumberton and Princeville, in North Carolina, were chosen for this study. The city of Lumberton is located in Robeson County on the Lumber River, and the Town of Princeville is located along the Tar River in Edgecombe County. These locations have been highly affected by several flood events, including Hurricane Matthew and Hurricane Florence.

The data used for the research include the following:

- UAV high-resolution imagery collected during hurricane Matthew (2016) over the study areas. These data were acquired by the North Carolina Emergency Management (NCEM). The size of each image is 4,000 x 4,000, with a resolution of 2.6 cm.
- The pre-flood LiDAR data acquired over the town of Princeville. This Quality Level 2 (QL2) LiDAR with two pulses per square meter (pls/m2) with an accuracy of 9.25-cm RMSE was collected by the North Carolina Emergency Management in 2014. This data was used to estimate the water level measurements.

3. METHODS AND DATA PROCESSING

3.1 Creating 3D flood map using SfM

SfM is a photogrammetric method commonly used for creating 3D models of a feature from overlapping 2D photographs taken from many locations and orientations. SfM works based on stereoscopic photogrammetry principles. That means triangulation is needed to calculate the 3D positions (x,y,z,) of objects from stereo pairs. In this study, we generated a 3D map from 80% overlapping UAV images. First, the pixels in the input images automatically identified matching features to create a point cloud model. These features can be corners or line segments and tracked from image to image to estimates the camera locations and orientations and the coordinates of the features. Based on the estimated camera positions, the locations of those points calculated and visualized as a 3D point cloud, in which each matchable pixels got its XYZ location in 3D space. Second, each set of three adjacent points connected into a triangular face to create a polygonal mesh model. Finally, the original images combined into a texture map and wrapped around the mesh to get a photo-realistic model.

3.2 Flooded Area Extraction Using FCN-8s

The second stage of the method involves extracting a 2D flood extent to classify the point cloud extracted from the SFM to water and non-water. For this task, a fully convolutional neural network (FCN) with a stride of 8 (FCN-8s) model was fine-tuned. The FCN-8s is composed of locally connected layers, such as convolution, pooling, and upsampling, without having any dense layer. This allows reducing the number of parameters and computation time. Given that all connections are local, FCN-8s can work on any image size. The FCN was proposed by Long et al. (2015) to train an end-to-end for semantic segmentation. In these models, VGG16 fully connected based classification layers were replaced by convolutional layers to maintain the 2-D structure of images. VGG-16 is a CNN architecture proposed by Simonyan et al. (2014) to investigate the effect of the convolutional network depth on its accuracy in the large-scale image recognition setting. The network is sketched in Figure 1.

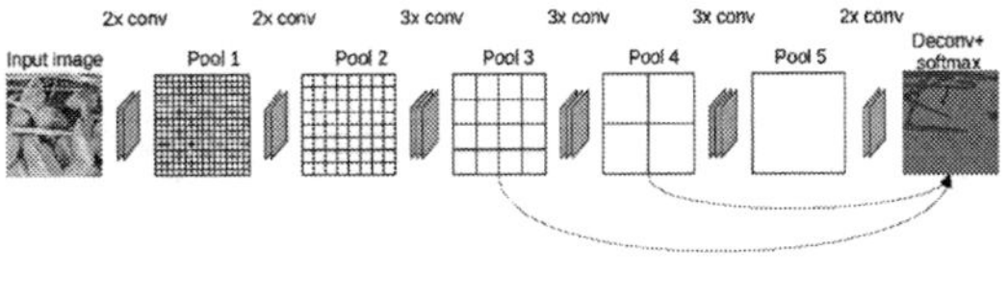

Figure 1. FCN-8s architecture (Skovsen et al., 2017)

To extract the 2D flooded area, the method consisted of annotating, training, classification, and evaluation stages. In the annotating stage, 150 UAV images were labeled manually using a MATLAB labeler tool. The network is trained using Stochastic Gradient Descent (SGD) for 6 epochs with a learning rate of 0.001, and a maximum batch size of 2. During training, 512- by-512 pixels size of 32 patches were randomly cropped and rotated in each batch size to increase the diversity of the training samples. We used a 10-fold cross-validation technique to evaluate the potential of the FCN-8s model. The importance of using 10-fold cross-validation is to give a less biased estimate of the FCN-8s model on unseen data. The training stage ended after 230,000 iterations for the experiments. In the classification stage, the performance of the network was tested using the unseen testing images. In the accuracy assessment stage, a confusion matrix was calculated to analyze the performance of the FCN-8s.

4. RESULTS AND DISCUSSION

4.1 SfM and FCN-8s Results

The FCN-8s approach for 2D flood mapping was implemented in the MATLAB 2019b, and Agisoft Methashape and ARCGIS software were used for creating cloud point using SFM and implemented an integrated approach. The computer was configured with 32 GB memory, an Intel(R) Xeon(R) ES-2620 v3 @ 2.40GHz ×2 processors memory, and a single NVIDIA Quadro M4000 GPU.

https://doi.org/10.5194/isprs-archives-XLIV-M-2-2020-25-2020

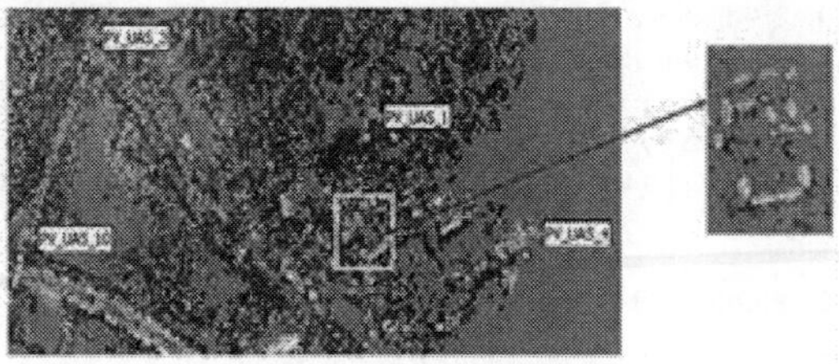

Figure 2. Point cloud model created by SfM

Our test area in Princeville (yellow box in Figure 2) did not contain any GCPs for georeferencing purposes. Thus, the images of larger areas were processed (3000 UAV images) for using the available GCPs in that area in the triangulation and image-to-image georeferencing. All analyses were done only for the study area.

Figure 3. (a) Orthophoto; (b) SfM based 3D flood map (side view); (c) SfM based 3D flood map (top view); (d) FCN-8s based 2D flood map

Figure 3d shows the 2D flood extent extracted by FCN-8s. The FCN-8s based classification achieved an accuracy of over 97% on distinguishing flooded areas from non-flooded. Once the FCN-8s model fine-tuned and trained with the dataset, the average processing time to extracted flooded areas was less than 2 seconds for a 4000x4000 pixels UAV image.

4.2 UAV and Lidar based DEMs results

The DEMs of the study area were generated using UAV data (figure 4a) and the preflooded Lidar data (figure 4b) to estimate the flood water level.

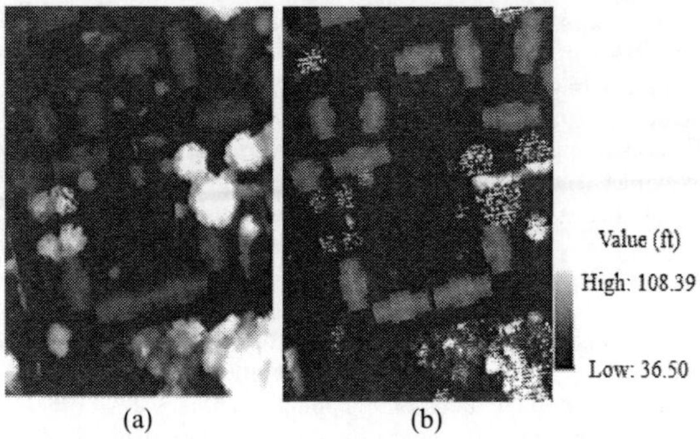

Figure 4. (a) UAV-based DSM (b) Lidar-based DSM

4.3 An Integerated 3D Classfication Result

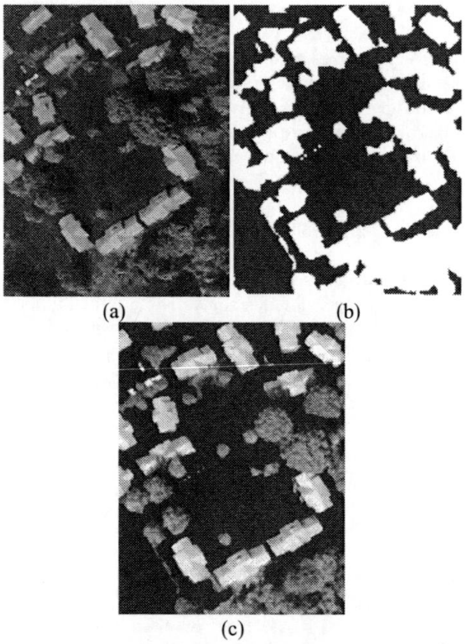

Figure 5. (a) 3D flood map (SfM); (b) FCN-8s extraction result; (c) Overlaid result

As it is shown in Figure 5a, it is difficult to identify the flooded areas from the SfM based 3D map directly. Figure 5c shows the result of the integrated approach (deep learning and SfM) obtained by overlaying the deep learning-based flood areas on the SFM-based DEM. The result shows precisely the flooded areas. The depth of the floodwater estimated by subtracting the UAV and LiDAR based DEMs. The DEM created using LiDAR considered as a benchmark to estimate the floodwater depth since the LiDAR data was acquired before the flood event. The minimum and maximum floodwater depth values calculated in the flooded areas are 0.32 ft and 1.24 ft, respectively. The estimated floodwater depth values give additional information that can be used for rescue and damage assessment tasks.

5. CONCLUSION

Flooding is a severe natural disaster that causes economic damage and loss of life and property. Accurate 3D mapping during

https://doi.org/10.5194/isprs-archives-XLIV-M-2-2020-25-2020

flooding events is vital for effective planning and rescue activities. The research proposed a method to create a 3D flood map by integrating SfM and deep learning. The results indicated that the deep learning-based method, such as FCN-8s, could accurately extract flooded areas with a classification accuracy of 97%. By overlaying the deep learning-based 2D flood map on the SfM-based DEM provides water level information. The 3D flood extent map can be used to support emergency response, and recovery activates during a flood event.

ACKNOWLEDGEMENT

The authors would like to thank the North Carolina Emergency Management for providing the UAV data, GCPs, and support data for the project. This material is based upon work supported by the National Science Foundation under Grants No.1800768 and No. 1832110, and North Carolina Collaboratory grant.

REFERENCES

Boccardo, P., Chiabrando, F., Dutto, F., Tonolo, F., Lingua, A.,2015. UAV deployment exercise for mapping purposes: Evaluation of emergency response applications. *Sensors*, 15(7), 15717–15737.

Caroti G, Martínez-Espejo Zaragoza I, Piemonte A., 2015a. Range and image-based modeling: a way for frescoed vault texturing optimization. *Int Arch Photogramm Remote Sens Spat Inf.* Sci. XL-5/W4, 285–290.

Ciregan, D., Meier, U., Schmidhuber, J., 2012. Multi-column deep neural networks for image classification. *In Proceedings of the IEEE Conference on Computer Vision and Pattern Recognition* (CVPR '12), IEEE, 3642–3649.

Farabet, C., Couprie, C., Najman, L., LeCun, Y., 2012. Learning hierarchical features for scene labeling. *IEEE transactions on pattern analysis and machine intelligence*, *35*(8), 1915-1929.

Feng, Q., Liu, J., Gong, J., 2015. Urban flood mapping based on unmanned aerial vehicle remote sensing and random forest classifier—A case of Yuyao, China. *Water*, *7*(4), 1437-1455.

Gebrehiwot, A., Hashemi-Beni, L., Thompson, G., Kordjamshidi, P., Langan, T. E., 2019. Deep convolutional neural network for flood extent mapping using unmanned aerial vehicle data. *Sensors*, *19*(7), 1486.

Girshick, R., Donahue, J., Darrell, T., Malik, J., 2014. Rich feature hierarchies for accurate object detection and semantic segmentation. In *Proceedings of the IEEE conference on computer vision and pattern recognition*, 580-587

Hashemi-Beni, L., Jones, J., Thompson, G., Johnson, C., Gebrehiwot, A., 2018. Challenges and opportunities for UAV-based digital elevation model generation for flood-risk management: A case of Princeville, North Carolina. *Sensors*, *18*(11), 3843.

Huang, X., Wang, C., Li, Z., 2018. A near real-time flood-mapping approach by integrating social media and post-event satellite imagery. *Annals of GIS*, *24*(2), 113-123.

Long, J., Shelhamer, E., Darrell, T., 2015. Fully convolutional networks for semantic segmentation. In *Proceedings of the IEEE conference on computer vision and pattern recognition*, 3431-3440.

Meesuk, V., Vojinovic, Z., Mynett, A. E., 2012. Using multidimensional views of photographs for flood modelling. In *2012 IEEE 6th International Conference on Information and Automation for Sustainability*, IEEE, 19-24.

Meesuk, V., Vojinovic, Z., Mynett, A. E., Abdullah, A. F., 2015. Urban flood modeling combining top-view LiDAR data with ground-view SfM observations. *Advances in Water Resources*, 75, 105-117.

Miller, C. L., Laflamme, R. A., 1958. The Digital Terrain Model-: *Theory & Application*. MIT Photogrammetry Laboratory.

Remondino, F., Spera, M. G., Nocerino, E., Menna, F., Nex, F., 2014. State of the art in high density image matching. *The photogrammetric record*, *29*(146), 144-166.

Sarker, C., Mejias, L., Maire, F., Woodley, A., 2019. Flood mapping with convolutional neural networks using spatio-contextual pixel information. *Remote Sensing*, *11*(19), 2331.
Simonyan, K., Zisserman, A., 2014. Very deep convolutional networks for large-scale image recognition. *arXiv preprint arXiv:1409.1556*.

Simonyan, K., Zisserman, A., 2014. Very deep convolutional networks for large-scale image recognition. *arXiv preprint arXiv:1409.1556*.

Skovsen, S., Dyrmann, M., Mortensen, A. K., Steen, K. A., Green, O., Eriksen, J., Karstoft, H., 2017. Estimation of the botanical composition of clover-grass leys from RGB images using data simulation and fully convolutional neural networks. *Sensors*, *17*(12), 2930.

Taigman, Y., Yang, M., Ranzato, M. A., Wolf, L., 2014. Deepface: Closing the gap to human-level performance in face verification. In *Proceedings of the IEEE conference on computer vision and pattern recognition*, 1701-1708.

Watts, A. C., Ambrosia, V. G., Hinkley, E. A., 2012. Unmanned aircraft systems in remote sensing and scientific research: Classification and considerations of use. *Remote Sensing*, *4*(6), 1671-1692.

EVALUATION OF CONVERTING LANDSAT DN TO TA AND SR VALUES ON SELECT SPECTRAL INDICES

A. L. Gettinger [1], R. Sivanpillai [2]

[1] Dept. of Ecosystem Science and Management, University of Wyoming, Laramie, WY 82071, USA, ORCID: 0000-0003-0664-2621 - agettin1@uwyo.edu
[2] Wyoming GIS Center, University of Wyoming, Laramie, WY 82071, USA, ORCID: 0000-0003-3547-9464 - sivan@uwyo.edu

KEY WORDS: Pre-processing, Digital number, ToA, Surface reflectance, Spectral indices, Landsat

ABSTRACT:

The complete archive of images collected across all Landsat missions has been reprocessed and categorized by the U.S. Geological Survey (USGS) into a three-tiered architecture: Real-time, Tier-1, and Tier-2. This tiered architecture ensures data compatibility and is convenient for acquiring high quality scenes for pixel-by-pixel change analyses. However, it is important to evaluate the effects of converting older Landsat images from digital numbers (DN) to top-of-the-atmosphere (TA) and surface reflectance (SR) values that are equivalent to more recent Landsat data. This study evaluated the effects of this conversion on spectral indices derived from Tier-1 (the highest quality) Landsat 5 and 8 scenes collected in 30 m spatial resolution. Spectral brightness and reflectance of mixed conifers, Northern Mixed Grass Prairie, deep water, shallow water, and edge water were extracted as DNs, TA, and SR values, respectively. Spectral indices were estimated and compared to determine if the analysis of these land cover classes or their conditions would differ depending on which preprocessed image type was used (DN, TA, or SR). Results from this study will be informative for others making use of indices with images from multiple Landsat satellites as well as for engineers planning to re-process images for future Landsat collections. This time-series study showed that there was a significant difference between index values derived from three levels of pre-processing. Average index values of vegetation cover classes were consistently significantly different between levels of pre-processing whereas average water index values showed inconsistent significant differences between pre-processing levels.

1. INTRODUCTION

1.1 Landsat Overview and Remote Sensing Importance

Landsat satellites have been collecting images since 1972 and are administered by the United States Geological Survey (USGS) in conjunction with the National Aeronautics and Space Administration (NASA) (Straub et al., 2019). These satellite images are provided at no-cost (USGS, 2018) and are utilized for a wide variety of applications including improving water use in vineyards (Ecker, 2020), wildfire and clear-cut mapping (Schroeder et al., 2011), invasive species mapping (Evangelista et al., 2009), and evaluating flooded areas (Sivanpillai et al., 2020, Wang et al., 2002).

1.2 Pre-processing

Prior to an image being used, pre-processing is often performed (Jiang et al., 2018) to try to reduce the effects from the atmosphere, the sun, topography, and the sensor itself (Young et al., 2017). Operations that are performed before the main evaluation of the data are considered pre-processing (Campbell, Wynne, 2011). The majority of pre-processing operations fall into the categories of geometric, absolute, and relative pre-processing (Young et al., 2017). Geometric pre-processing is performed to achieve accurate image location and includes image orthorectification and georeferencing (Young et al., 2017). Adjustments for effects of the sun, topography, the atmosphere, and the sensor are included in absolute pre-processing, with absolute pre-processing aiding in values being able to be evaluated with other values of the same amount of pre-processing that were acquired from different sensors, times, or locations (Young et al., 2017). Relative pre-processing is performed to bring about radiometric scale equivalence between matching bands of a reference image and an evaluated image (Young et al., 2017).

1.3 Data Types

Light that is reflected from features on the Earth's surface is registered by a satellite scanner as brightness, or radiance (Campbell and Wynne, 2011, Young et al., 2017), in W • micrometer^{-1} • steradian^{-1} • m^{-2} (Campbell and Wynne, 2011). The analog signals from the sensor are turned into numbers (Ose et al., 2016). A pixel in a digital image stores the brightness data as numeric values that are called digital numbers (DNs) (Campbell and Wynne, 2011). Digital numbers are noted as binary bits that function as a scale denoting approximate brightness (Campbell and Wynne, 2011). The number of bits determines how many values can be registered for each sensor band and the overall image (Campbell and Wynne, 2011). Data across the Landsat sensors vary in bits, with data from MSS being 6- and 7-bit, TM and ETM+ being 8-bit, and OLI being 12-bit (USGS, 2019b).

Since digital numbers do not give radiance in physical units (Campbell and Wynne, 2011), digital numbers cannot be used to evaluate spectral values through time (Young et al., 2017), and digital numbers cannot be used to evaluate brightness in physical units from one scene to another (Campbell and Wynne, 2011). However, the metadata files of satellite images contain calibration coefficients that can be utilized to obtain absolute radiance from digital numbers; this radiance is commonly called at-sensor radiance and is one of the ways that an image can be pre-processed (Young et al., 2017).

This contribution has been peer-reviewed.
https://doi.org/10.5194/isprs-archives-XLIV-M-2-2020-29-2020

Additional absolute pre-processing can be performed beyond at-sensor radiance (Young et al., 2017). The sun's angle of elevation, length between the Sun and the Earth, and exoatmospheric irradiance from the sun can be addressed by calculating top-of-the-atmosphere reflectance (TA), which gives the amount of reflected radiation that is being received by a sensor at the top of the atmosphere (Young et al., 2017).

Converting TA data to surface reflectance aims to minimize how electromagnetic radiation is impacted by particles such as atmospheric aerosols, water vapor, and gases (Young et al., 2017). The radiative transfer models used in calculating surface reflectance for Landsat 4-5, and 7 is the 6S method (Vermote et al., 1997, USGS, 2019a), whereas Landsat 8 uses an internal algorithm (USGS, 2019a).

1.4 Landsat Data Product Structure and Processing

Data products from images procured from Landsat missions 1-8 are available in Landsat Collection 1 (USGS, 2018). Within Landsat Collection 1, there are three tiers (USGS, 2018). Depending on the amount of processing performed and quality, data products are designated as Real-Time (RT), Tier 1 (T1), or Tier 2 (T2) (USGS, 2018). After receiving preliminary processing, data from Landsat 8 OLI/TIRS and Landsat 7 ETM+ are placed into the Real-Time tier; this data is subsequently processed again and moved into Tier 1 or Tier 2 (USGS, 2018). Products in Tier 1 fulfill specific quality requirements; Tier 2 products are lower in quality compared to products in Tier 1 (USGS, 2018).

During Level 1 Processing, MSS data are scaled from 6- and 7-bit data to 8-bit, and OLI data are scaled from 12-bit to 16-bit (USGS, 2019b). Additional processing information can be found in the Landsat Collection 1 Level 1 Product Definition (USGS, 2019b).

The possibility that pre-processing could cause artifacts or errors has been noted (Campbell and Wynne, 2011, Young et al., 2017). Roy et al. (2016) has also evaluated Normalized Difference Vegetation Index (NDVI) values between Landsat 7 ETM+ and Landsat 8 OLI for both surface reflectance and top of the atmosphere. However, there is a lack of studies evaluating the effects of the conversion of digital numbers to top-of-the-atmosphere and surface reflectance values on spectral indices.

Given the wide variety of practical applications for which Landsat images are used, understanding if there are effects in derived products from the underlying image pre-processing is needed. This study evaluated the conversion of digital numbers to top-of-the-atmosphere and surface reflectance values on select spectral indices calculated for five cover classes.

2. MATERIALS AND METHODS

2.1 Image Acquisition and Pre-processing

Four Landsat 5 Thematic Mapper (TM) and two Landsat 8 Operational Land Imager (OLI) scenes were downloaded from Landsat Collection 1 from USGS Earth Explorer. Scenes were acquired in 1998, 2000, 2002, 2004, 2014, and 2016. Each of the six scenes was downloaded as compressed files in three pre-processing levels: digital number (DN), top-of-the-atmosphere (TA), and surface reflectance (SR) for a total of eighteen scenes. From each compressed file, the three visible and three infrared bands were extracted and stacked for a total of six bands.

2.2 Study Site Description

Spectral reflectance and DN values were extracted from these images for the following classes: predominately mixed conifers (MC), predominately Northern Mixed Grass Prairie (NMGP), deep water (DW), shallow water (SW), and edge water (EW).

2.2.1 Predominately Mixed Conifers: The MC cover class pixels were extracted from an area on National Forest land within the Black Hills of South Dakota (SD). The main tree species in the Black Hills is the ponderosa pine (*Pinus ponderosa*), but also includes other tree species such as Rocky Mountain juniper (*Juniperus scopulorum*), quaking aspen (*Populus tremuloides*), white spruce (*Picea glauca*), and bur oak (*Quercus macrocarpa*) (Walters et al., 2013). Mean annual precipitation in the nearby town of Custer, SD is 49.91 centimeters (cm) and mean annual temperature is 6.83°C (44.3°F) (Arguez et al., 2010).

An area to the southwest of Jewel Cave National Monument [43.729832, -103.829320] was selected for its relative homogeneity in canopy cover and size. These sites were initially identified with Google Maps and later confirmed through field visits. From visual inspection, the stands appeared to be dominated by ponderosa pine with interspersed Rocky Mountain juniper. Due to the mixture of conifer species and potential for understory vegetation to also be a part of the spectral reflectance, the cover class was designated as predominately MC.

Additionally, the closest year available in Google Earth Pro was used to evaluate historic ground conditions and to ensure uniformity in canopy cover and tree condition (live or dead) across time for each sampling location. Google Earth Pro was also used to establish approximate 90 x 90-meter (m) plots for each image of the closest year to the corresponding Landsat image. Geographic coordinates were recorded from each plot in ERDAS Imagine using a combination of Google Earth Pro and Google Maps.

2.2.2 Predominately Northern Mixed Grass Prairie: For the predominately NMGP class, an area of land in northeastern Wyoming (WY) was chosen approximately five miles north of the town of Newcastle, WY [43.854969, -104.206915]. This area was selected due to its familiarity, and site location was determined by viewing Google Maps for size and relative homogeneity in cover. While ground inspection of the site was not performed, major species in the northern mixed grass prairie of the United States are needle-and-thread (*Hesperostipa comata*), western wheatgrass (*Pascopyrum smithii*), and blue grama (*Bouteloua gracilis*) (Singh et al., 1983).

The closest year available in Google Earth Pro was also used for the NMGP cover class to evaluate historic ground conditions and try to ensure the most uniformity in grass ground cover for each sampling location. The cover class was designated as predominately mixed NMGP to account for additional variables, such as bare ground or rock. Google Earth Pro was again used to establish approximate 90 x 90-meter (m) plots for each image of the closest year to the corresponding Landsat image. Geographic coordinates were recorded from each plot in ERDAS Imagine using a combination of Google Earth Pro and Google Maps.

https://doi.org/10.5194/isprs-archives-XLIV-M-2-2020-29-2020 | © Authors 2020. CC BY 4.0 License.

2.3 Water Cover Classes: For the three water classes, Keyhole Reservoir [44.360154, -104.850673] in northeastern Wyoming was selected. A bathymetric map of Keyhole (Ferrari, 2005) was used for ocular selection of areas that were shallow and deep water, with shallow water being designated as approximately 1240 m (4070 feet) to 1245 m (4085 feet) in elevation on the map. Deep water was considered to fall between approximately 1228 m (4030 feet) to 1231 m (4040 feet) in elevation on the map. Pixels that appeared along the perimeter of the waterbody and possessed mixed coloring between the waterbody and the land were visually selected as edge water. Geographic coordinates were recorded from ERDAS Imagine for each sample.

2.4 Data collection

File pixel values were extracted using the 'Inquire' function in ERDAS Imagine for the five cover classes. For every year, the three pre-processed images (DN, TA, SR) were displayed in the ERDAS viewer and file pixel values were collected from the same latitude and longitude for each pre-processed image. The number of samples per cover class varied, with numbers ranging from seven to fourteen depending on site suitability. Additional samples were limited for certain years due to variables including cloud cover, differences in vegetation cover, canopy closure, or tree health.

2.5 Spectral Indices

Spectral indices were calculated from the file pixel values for each product level across time. Select water and vegetation indices are given in Table 1.

Cover Class	Index	Formula	Reference
MC NMGP EW	NDVI	NIR-Red/ NIR+Red	Rouse et al., 1973
MC NMGP EW	Simple Ratio	Red/ NIR	Pearson and Miller, 1972†
MC NMGP EW	Vegetation Index	NIR/R	Lillesand and Kiefer, 1987† Jordan, 1969
DW SW EW	NDWI	Green-NIR/ Green+NIR	McFeeters, 1996
DW SW EW	MNDWI	Green-SWIR/ Green+SWIR	Xu, 2006
DW SW EW	WII*	NIR^2/ Red	Caillaud et al., 1991†
DW SW EW	MWII**	SWIR^2/ Red	Davrache et al., 2013
DW SW EW	WI***	NIR^2/ Green	Davrache et al., 2013
DW SW EW	MWI****	SWIR^2/ Green	Davrache et al., 2013

Table 1. Spectral indices calculated for a given cover class. NOTES: *Water Impoundment Index; **Modified Water Impoundment Index; ***Water Index; ****Modified Water Index; †As cited in Davranche et al., 2013

Both vegetation and water indices were calculated for the edge water pixels due to the potential for vegetation to be incorporated in the pixel. Formula for these indices were adjusted to accommodate the coastal blue and cirrus bands of OLI.

Paired, two-tailed t-tests were calculated between the different product levels for each index to determine the significance between index averages. Since this experiment was evaluating recurrent measurements from the same groups (MC, NMGP, DW, SW, and EW), standard error bars are not applicable for determining significance (Cumming et al., 2007). Hence difference in averages were compared at an alpha value of 0.05.

3. RESULTS

3.1 Vegetation

All three vegetation indices computed for both vegetation cover classes (predominately mixed conifer and predominately Northern Mixed Grass Prairie) exhibited a significant difference ($p<0.001$) between each pre-processing level for all six years.

3.1.1 Predominately Mixed Conifers: Index values from all three pre-processing levels calculated from reflectance of MC pixels were consistently different from each other. There was a significant difference ($p <0.001$) between each pre-processing level for the three indices calculated across all six years (Fig. 1).

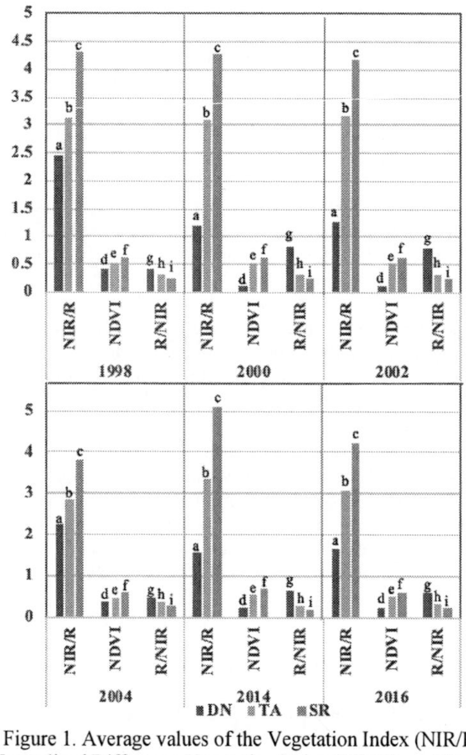

Figure 1. Average values of the Vegetation Index (NIR/R), Normalized Difference Vegetation Index (NDVI), and Simple Ratio Index (R/NIR) for predominately mixed conifer at three pre-processing levels (DN, TA, SR) across six years. Letters denote significant difference ($p<0.001$) between pre-processing levels within a given index. Average index values for pre-processing levels were not compared across different indices.

3.1.2 Predominately Northern Mixed Grass Prairie: Index values calculated from reflectance of NMGP pixels were also consistently different between data products, with there being a significant difference ($p < 0.001$) between each pre-processing level for the three indices calculated across all six years (Fig. 2).

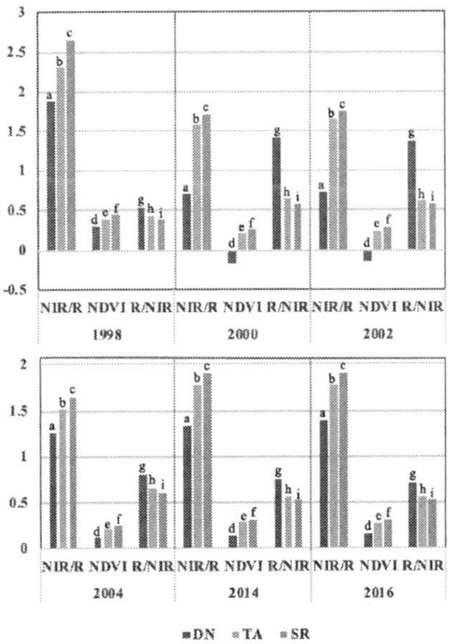

Figure 2. Average values of the Vegetation Index (NIR/R), Normalized Difference Vegetation Index (NDVI), and Simple Ratio Index (R/NIR) for predominately Northern Mixed Grass Prairie at three pre-processing levels (DN, TA, SR) across six years. Letters denote significant difference ($p<0.001$) between pre-processing levels within a given index. Average index values for pre-processing levels were not compared across different indices.

3.2 Water

Average index values exhibited less consistency in significant difference between the three levels of pre-processing. Deep water exhibited the highest consistency of significant difference ($p<0.05$ for 106/108 p-values), with shallow water exhibiting lower consistency ($p<0.05$ for 101/108 p-values), and edge water pixels exhibiting the least amount of consistency in significant difference between the three levels of pre-processing ($p<0.05$ for 133/162 p-values). The edge water had a greater number of p-values calculated in comparison to deep water or edge water due to the three vegetation indices that were also computed for edge water.

3.2.1 Deep Water: There was a significant difference ($p < 0.05$) between each pre-processing level for the three indices calculated across all six years for DW pixels with only two exceptions (Fig. 3, Fig. 4). For the NDWI index calculated from a scene collected on July 6th, 1998 scene, there was not a significant difference between values calculated from digital number and surface reflectance products ($p = 0.71$) (Fig. 3).
The MNDWI index calculated for a scene collected on June 23rd, 2002 also exhibited a non-significant difference between values calculated from digital number and surface reflectance products ($p = 0.10$) (Fig. 3).

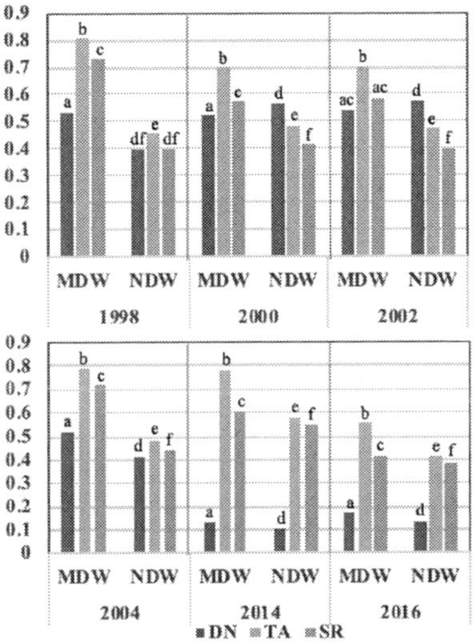

Figure 3. Average values of the Modified Normalized Difference Water Index (abbreviated MDW in the graph) and the Normalized Difference Water Index (abbreviated NDW in the graph) of deep water for three pre-processing levels (DN, TA, SR) across six years. Letters denote significant difference ($p < 0.05$) between pre-processing levels within a given index. Average index values for pre-processing levels were not compared across different indices.

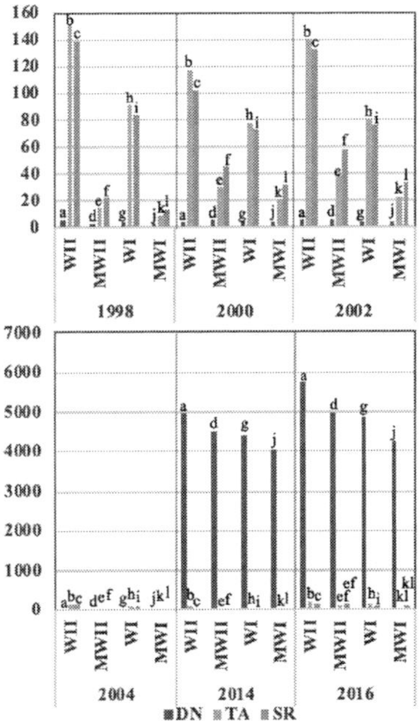

Figure 4. Average values of the Water Impoundment Index (WII), Modified Water Impoundment Index (MWII), Water Index (WI), and Modified Water Index (MWI) of deep water for three pre-processing levels (DN, TA, SR) across six years. Letters denote significant difference ($p < 0.05$) between pre-processing levels within a given index. Average index values for pre-processing levels were not compared across different indices.

3.2.2 Shallow Water: Index values calculated from reflectance of SW pixels exhibited less consistency in significant difference across data products in comparison to DW and vegetation cover classes. The majority (101/108) of p-values between data products across the six years were significant ($p<0.05$) (Fig. 5, 6). Four years had one insignificant p-value and one year had three insignificant p-values (Table 2, Fig. 5, 6). In the year 2000, all p-values were significant (Fig. 5, 6). Four of the seven insignificant p-values were calculated from the NDWI index, with one p-value being from the DN-SR comparison, and the other three values resulting from the TA-SR comparison (Table 2, Fig. 5).

Year	Index	Pre-processing Levels Compared		
		DN-TA	DN-SR	TA-SR
7/6/98	*NDWI*	1.43E-09	**1.50E-01**	3.00E-02
6/23/02	*NDWI*	3.31E-07	1.20E-05	**9.03E-02**
8/7/04	*NDWI*	4.76E-11	1.26E-06	**2.81E-01**
9/20/14	*NDWI*	1.37E-11	2.54E-08	**7.29E-01**
6/21/16	*MNDWI*	6.79E-10	5.89E-08	**9.79E-02**
6/21/16	*MWII*	2.29E-09	2.07E-09	**1.85E-01**
6/21/16	*MWI*	2.79E-09	2.60E-09	**1.18E-01**

Table 2. P-values from years that exhibited an insignificant difference ($p>0.05$) between select indices calculated from DN, TA, and SR pre-processing levels for shallow water. Bold numbers denote insignificant p-values.

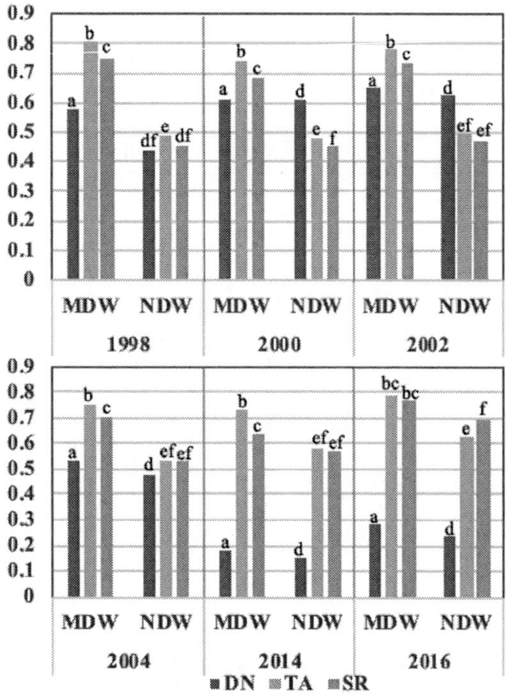

Figure 5. Average values of the Modified Normalized Difference Water Index (abbreviated MDW in the graph) and the Normalized Difference Water Index (abbreviated NDW in the graph) of shallow water for three pre-processing levels (DN, TA, SR) across six years. Letters denote significant difference ($p<0.05$) between pre-processing levels within a given index.

Average index values for pre-processing levels were not compared across different indices.

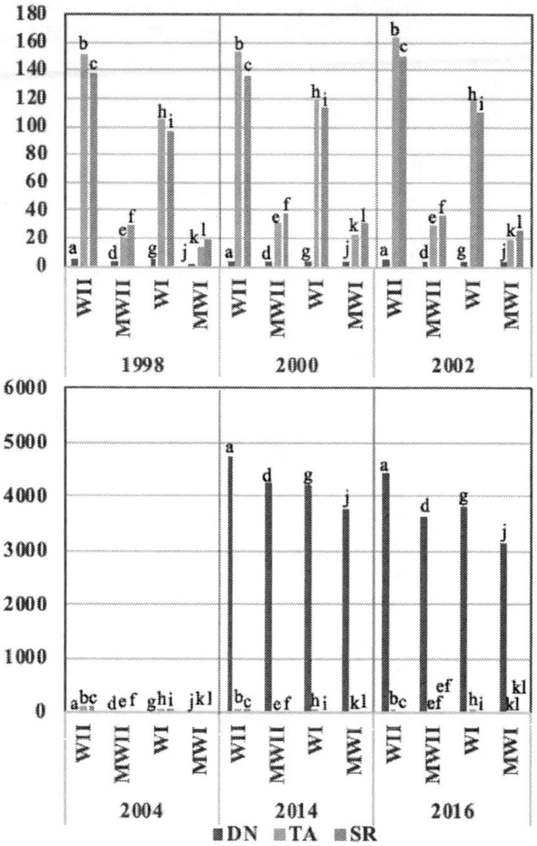

Figure 6. Average values of the Water Impoundment Index (WII), Modified Water Impoundment Index (MWII), Water Index (WI), and Modified Water Index (MWI) of shallow water for three pre-processing levels (DN, TA, SR) across six years. Letters denote significant difference ($p<0.05$) between pre-processing levels within a given index. Average index values for pre-processing levels were not compared across different indices.

3.2.3 Edge Water: The EW pixels exhibited the most inconsistency in significant difference between the three levels of pre-processing. The majority of average index values exhibited a significant difference ($p<0.05$) between the three levels of pre-processing (Fig. 7, 8, 9). However, out of 162 average index values calculated, there were 29 instances where there was not a significant difference ($p>0.05$) between levels of pre-processing (Table 3).

The majority (27/29) of insignificant values between the three levels of pre-processing came from the indices calculated from scenes acquired by Landsat 8 OLI in years 2014 and 2016 (13 and 14 insignificant values, respectively) (Table 3, Fig. 7, 8, 9). The NDWI had the most insignificant values (4/29) out of all the indices (Table 4, Fig. 8).

https://doi.org/10.5194/isprs-archives-XLIV-M-2-2020-29-2020 | © Authors 2020. CC BY 4.0 License.

Year	Index	Pre-processing Levels Compared		
		DN-TA	DN-SR	TA-SR
7/6/98	*NDWI*	**3.15E-01**	8.43E-06	4.05E-06
8/7/04	*NDWI*	**3.23E-01**	4.47E-05	6.68E-07
9/20/14	*NIR/R*	**5.06E-01**	**2.33E-01**	3.07E-02
9/20/14	*NDVI*	**6.71E-01**	**8.39E-01**	7.09E-03
9/20/14	*MNDWI*	**2.00E-01**	**7.24E-01**	9.20E-07
9/20/14	*NDWI*	**2.59E-01**	**9.33E-01**	7.97E-04
9/20/14	*WII*	7.95E-07	2.62E-07	**1.67E-01**
9/20/14	*WI*	1.94E-06	3.48E-07	**1.13E-01**
9/20/14	*SR*	**2.25E-01**	**3.76E-01**	**7.16E-02**
6/21/16	*NIR/R*	**3.36E-01**	**8.23E-01**	**9.47E-02**
6/21/16	*NDVI*	**7.47E-02**	**2.05E-01**	**5.50E-01**
6/21/16	*MNDWI*	1.46E-02	**2.05E-01**	2.24E-04
6/21/16	*NDWI*	1.83E-02	**1.57E-01**	**7.21E-02**
6/21/16	*WII*	3.00E-05	2.26E-05	**6.55E-01**
6/21/16	*WI*	6.47E-05	4.00E-05	**2.46E-01**
6/21/16	*SR*	**5.61E-02**	**1.20E-01**	**3.94E-01**

Table 3. P-values from years that exhibited an insignificant difference (p >0.05) between select indices calculated from DN, TA, and SR pre-processing levels. Bold numbers denote insignificant p-values.

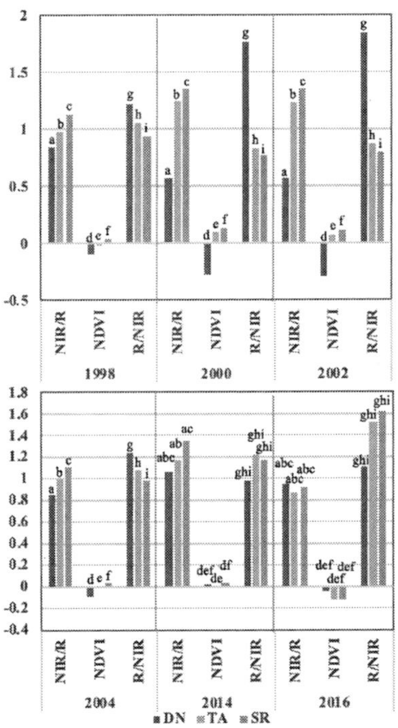

Figure 7. Average values of the Vegetation Index (NIR/R), Normalized Difference Vegetation Index (NDVI), and Simple Ratio Index (R/NIR) of the edge water class for three pre-processing levels (DN, TA, SR) across six years. Letters denote significant difference ($p<0.05$) between pre-processing levels within a given index. Average index values for pre-processing levels were not compared across different indices.

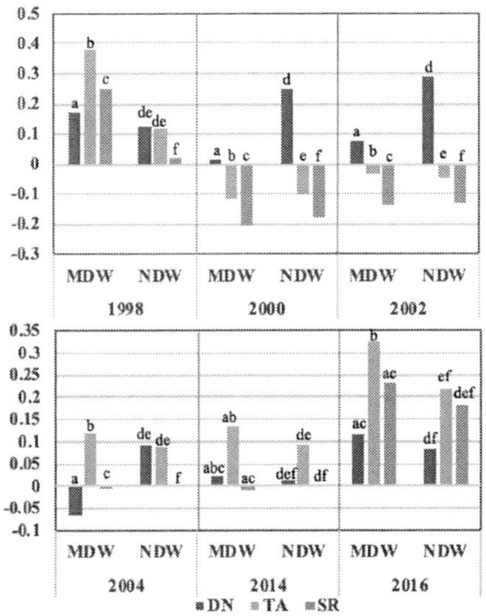

Figure 8. Modified Normalized Difference Water Index (MDW) and Normalized Difference Water Index (NDW) of the edge water class for three pre-processing levels (DN, TA, SR) across six years. Letters denote significant difference ($p<0.05$) between pre-processing levels within a given index. Average index values for pre-processing levels were not compared across different indices.

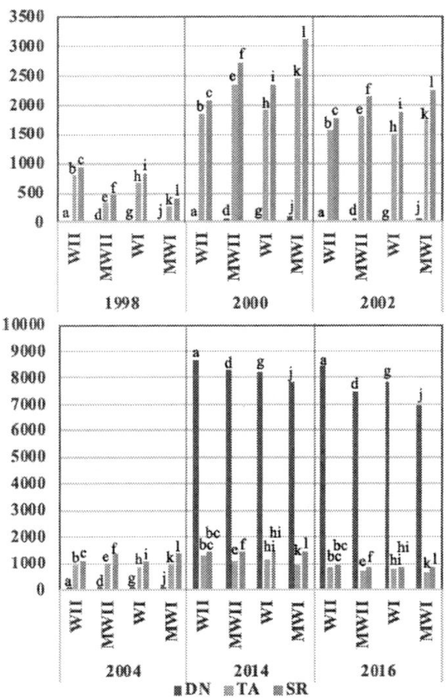

Figure 9. Average values of the Water Impoundment Index (WII), Modified Water Impoundment Index (MWII), Water Index (WI), and Modified Water Index (MWI) of edge water for three pre-processing levels (DN, TA, SR) across six years.

https://doi.org/10.5194/isprs-archives-XLIV-M-2-2020-29-2020 | © Authors 2020. CC BY 4.0 License.

Letters denote significant difference ($p<0.05$) between pre-processing levels within a given index. Average index values for pre-processing levels were not compared across different indices.

4. DISCUSSION AND CONCLUSIONS

4.1 Vegetation

There was a significant difference in average index values across year, sensor (Landsat 5 TM, Landsat 8 OLI), index, vegetation cover class (MC, NMGP), and pre-processing level (DN, TA, SR). Users need to be aware that different levels of pre-processing in vegetation could have higher or lower average index values due to pre-processing effects rather than due to vegetation characteristics and conditions. All average index values exhibited consistent, highly significant differences ($p<0.001$) between different levels of pre-processing. Therefore, individuals who are employing indices to create maps need to be careful when comparing products generated from different levels of pre-processing in order to make accurate conclusions about ground conditions. Further research into the effects of pre-processing on spectral indices using additional vegetation cover classes, more indices, and a larger sample size is needed.

4.2 Water

Overall, average index values from different levels of pre-processing for Landsat satellites 5 and 8 for all three water cover classes (DW, SW, EW) were not consistently affected by pre-processing. Significant difference in average index values varied by year, by sensor, by index, by water feature, and by pre-processing level (DN, TA, SR).

Pre-processing did not systematically affect average index values across all years. For some images, all average index values were significantly different whereas other images from other years had multiple average index values that were not significant from one another. Even across images from different years that were collected by the same sensor, significance between average index values of the three levels of pre-processing was not always consistent.

Moreover, data acquired by both sensors were not affected by pre-processing equally. Landsat 8 OLI data had less significant differences than did Landsat 5 TM data for both EW and SW. However, Landsat 5 data had fewer significant values for DW than Landsat 8 data. Average index values of edge water from Landsat 8 data had fewer significant differences between levels of pre-processing in comparison to index values of edge water from Landsat 5 data, or deep and shallow water from Landsat 8 or Landsat 5 data. Furthermore, the average index values of the WII, MWII, WI, and MWI calculated from the DN level of pre-processing were very high in comparison to the average index values calculated from TA and SR for scenes acquired by Landsat 8 in 2014 and 2016. This large difference in values from Landsat 8 may not only be due to the differences in pre-processing levels and requires further investigation.

Additionally, some index values were not consistently significantly different across all years. However, out of the six water indices that were calculated, MNDWI and NDWI had at least one insignificant difference between index values of the varying pre-processing levels for each water cover class. This lack of predictability in whether or not MNDWI or NDWI values significantly differ between pre-processing levels is important for users to be informed of as MNDWI and NDWI are widely used. Additional research should evaluate why specific bands that have been pre-processed to a certain level are resulting in average index values being significantly different from one another.

Pre-processing effects also appeared to decrease along a gradient that occurred with the lake's physical characteristics. Deep water was the most impacted by pre-processing and edge water was the least impacted by pre-processing effects. Shallow water was intermediately affected.

Furthermore, there was not a pre-processing level that consistently exhibited the highest or lowest average index value across all indices nor within an index for all years. Significant differences in the average index values between the pre-processing levels (DN-TA, DN-SR, TA-SR) was also not consistent across all years or indices. Further research to evaluate why a significant difference exists between average index values from DN-SR, but not between DN-TA or TA-SR is needed.

The lack of consistency in the significance between levels of pre-processing and the lack of consistency in size of average index values across the three pre-processing levels demonstrates the need for further research to be performed into the effects of pre-processing on spectral indices used with water. A larger sample size, additional water indices, and use of further satellites should be investigated.

Users of Landsat satellite data need to be aware of the effects of pre-processing on spectral indices calculated from water classes are not predictable. This lack of predictability also demonstrates the importance of specifying what level of pre-processing was used to generate derived products, such as maps, in addition to specifying what indices are being used. Additionally, users should be aware of what pre-processing level was used in the creation of threshold values for image classification. Due to the differences in index values within the same index for different levels of pre-processing, index values could be falling above or below the threshold as a result of pre-processing rather than actual ground conditions. Caution should be exhibited when trying to compare derived products that are generated using different indices or different pre-processing levels due to the current lack of predictability in the effects that pre-processing has across satellites and indices on waterbodies.

In conclusion, levels of pre-processing have varied effects on spectral index values calculated for vegetation and water features. Users need to exercise caution in selecting a level of pre-processing for their data or when comparing data products derived from different levels of pre-processing.

ACKNOWLEDGEMENTS

We would like to thank WyomingView for project support. Thank you to the United States Geological Survey for providing no-cost Landsat data and AmericaView. Thank you to Kay Gettinger for her support during this project and review assistance. The authors declare no conflict of interest.

This material is based upon work supported by the U.S. Geological Survey under Grant/Cooperative Agreement No. G18AP00077 to the second author. The views and conclusions contained in this document are those of the authors and should not be interpreted as representing the opinions or policies of the U.S. Geological Survey. Mention of trade names or commercial products does not constitute their endorsement by the U.S. Geological Survey. This manuscript is submitted for publication with the understanding that the United States Government is authorized to reproduce and distribute reprints for Governmental purposes.

REFERENCES

Arguez, A., Durre, I., Applequist, S., Squires, M., Vose, R., Yin, X., Bilotta, R., 2010a. NOAA's U.S. Climate Normals (1981-2010). [Annual/Seasonal Normals, Custer, SD US]. NOAA National Centers for Environmental Information DOI: 10.7289/V5PN93JP [July 13 2020].

Arguez, A., Durre, I., Applequist, S., Squires, M., Vose, R., Yin, X., Bilotta, R., 2010b. NOAA's U.S. Climate Normals (1981-2010). [Annual/Seasonal Normals, Newcastle, WY US]. NOAA National Centers for Environmental Information DOI: 10.7289/V5PN93JP [13 July 2020].

Campbell, J.B., Wynne, R.H., 2011. *Introduction to Remote Sensing*, 5th edition. The Guilford Press, NY.

Cumming, G., Fidler, F., Vaux, D.L., 2007. Error bars in experimental biology. *Journal of Cell Biology* 177(1), 7-11. DOI: 10.1083/jcb.200611141.

Davranche, A., Poulin, B., Lefebvre, G., 2013. Mapping flooding regimes in Camargue wetlands using seasonal multispectral data. *Remote Sensing of Environment* 138, 165-171. DOI: 10.1016/j.rse.2013.07.015.

Ecker, M, 2020. Raising a glass in wine country to better water management. landsat.gsfc.nasa.gov/raising-a-glass-in-wine-country-to-better-water-management/ (7 July 2020).

Evangelista, P.H., Stohlgren, T.J., Morisette, J.T., Kumar, S., 2009. Mapping invasive Tamarisk (*Tamarix*): A comparison of single-scene and time-series analyses of remotely sensed data. *Remote Sensing* 1 (3), 519-533. DOI: 10.3390/rs1030519.

Ferrari, 2005. Keyhole Reservoir 2003 Sedimentation Survey: United States Department of the Interior Bureau of Reclamation, 21 p.
www.usbr.gov/tsc/techreferences/reservoir.html (12 July 2020).

Xu, H., 2006. Modification of normalised difference water index (NDWI) to enhance open water features in remotely sensed imagery. *International Journal of Remote Sensing,* 27(14), 3025-3033. DOI: 10.1080/01431160600589179.

Jiang, W., He, G., Long, T., Ni, Y. Liu, H., Peng, Y., Lv, K., Wang, G., 2018. Multilayer perceptron neural network for surface water extraction in Landsat 8 OLI satellite images. *Remote Sensing.* 10(5), 755. DOI: 10.3390/rs10050755.

Jordan, C.F., 1969. Derivation of leaf-area index from quality of light on the forest floor. *Ecology* 50(4), 663-666. DOI: 10.2307/1936256.

McFeeters, S.K., 1996. The use of the Normalized Difference Water Index (NDWI) in the delineation of open water features. *International Journal of Remote Sens.* 17(7), 1425-1432. DOI: 10.1080/01431169608948714.

Ose, K., Corpetti, T., Demagistri, L., 2016. Multispectral satellite image processing. In Baghdadi, N., Zribi, M. (Ed.), *Optical Remote Sensing of Land Surface: Techniques and Methods*, 57-124. DOI: 10.1016/B978-1-78548-102-4.50002-8.

Rouse, J.W., Haas, R.H., Schell, J.A., Deering, D.W., 1973. Monitoring vegetation systems in the Great Plains with ERTS. NASA Goddard Space Flight Center 3d ERTS-1 Symp., Vol. 1, Sect. A., 309-317. ntrs.nasa.gov/search.jsp?R=19740022614 (13 July 2020).

Roy, D.P., Kovalskyy, V., Zhang, H.K., Vermote, E.F., Yan, L., Kumar, S.S., Egorov, A. 2016. Characterization of Landsat-7 to Landsat-8 reflective wavelength and normalized difference vegetation index continuity. *Remote Sensing of Environment* 185: 57-70. DOI: 10.1016/j.rse.2015.12.024.

Schroeder, T.A., Wulder, M.A., Healey, S.P., Moisen, G.G., 2011. Mapping wildfire and clearcut harvest disturbances in boreal forests with Landsat time series data. *Remote Sensing of Environment,* 115(6), 1421-1433. DOI: 10.1016/j.rse.2011.01.022.

Singh, J.S., Lauenroth, W.K., Heitschmidt, R.K., Dodd, J.L., 1983. Structural and functional attributes of the vegetation of northern mixed prairie of North America. *Botanical Review* 49(1), 117-149. DOI: 10.1007/BF02861010.

Sivanpillai, R., Jacobs, K.M., Mattilio, C.M., Piskorski, E.V., 2020. Rapid flood inundation mapping by differencing water indices from pre- and post-flood Landsat images. *Frontiers of Earth Sciences* DOI: 10.1007/s11707-020-0818-0.

Straub, C.L., Koontz, S.R., Loomis, J.B., 2019. Economic valuation of Landsat Imagery: USGS Open-File Report 2019-1112, 13 p. DOI: 10.3133/ofr20191112.

USGS, 2018. Landsat collections: USGS Fact Sheet 2018-3049, 2 p. DOI: 10.3133/fs20183049.

USGS, 2019a. Landsat 8 surface reflectance code (LASRC) product guide, LSDS- 1368, Version 2.0. prd-wret.s3.us-west-2.amazonaws.com/assets/palladium/production/atoms/files/LSDS-1368_L8_SurfaceReflectanceCode-LASRC_ProductGuide-v2.pdf (16 July 2020).

USGS, 2019b. Landsat Collection 1 Level 1 product definition, LSDS-1656, Version 2.0. prd-wret.s3-us-west-2.amazonaws.com/assets/palladium/production/atoms/files/LSDS-1656_%20Landsat_Collection1_L1_Product_Definition-v2.pdf (17 July 2020).

Vermote, E.F., Tanre, D., Deuze, J.L., Herman, M., Morcette, J.-J., 1997. Second simulation of the satellite signal in the solar spectrum, 6S: an overview. *Transactions on Geoscience and Remote Sensing*, 35(3), 675-686. DOI: 10.1109/36.581987.

Walters, B.F., Woodall, C.W., Piva, R.J., Hatfield, M.A., Domke, G.M., Haugen, D.E, 2013. Forests of the Black Hills National Forest 2011. United States Department of Agriculture, United States Forest Service Resource Bulletin NRS-83. www.nrs.fs.fed.us/pubs/rb/rb_nrs83.pdf (12 July 2020).

Young, N.E. Anderson, R.S., Chignell, S.M., Vorster, A.G. Lawrence, R., Evangelista, P.H. 2017. A survival guide to Landsat preprocessing. *Ecology* 98(4), 920-932. DOI: 10.1002/ecy.1730.

THE CASE FOR LOW-COST, PERSONALIZED VISUALIZATION FOR ENHANCING NATURAL HAZARD PREPAREDNESS

Peter Gmelch[1], Raul Lejano[1], Evan O'Keeffe[2], Debra F. Laefer[2], Cady Drell[1], Michela Bertolotto[3], Ulrich Ofterdinger[4], Jennifer McKinley[4]

[1] Steinhardt School of Culture, Education, and Human Development, East Bldg, 239 Greene Street, Floor 6, NY, NY - (plg275, lejano, ced478)@nyu.edu
[2] Center for Urban Science and Progress and the Department of Civil and Urban Engineering, Tandon School of Engineering, New York University, 370 Jay St., 1301C, Brooklyn, NY - (eo52, debra.laefer)@nyu.edu
[3] School of Computer Science, University College Dublin, Belfield, Dublin 4, Ireland - michela.bertolotto@ucd.ie
[4] School of Natural and Built Environment, Queen's University, Stranmillis Rd, Belfast, BT9 5AG UK - (u.ofterdinger, j.mckinley)@qub.ac.uk

KEY WORDS: Imagery, Augmented Reality, Virtual Reality, Flood, Risk, Communication

ABSTRACT:

Each year, lives are needlessly lost to floods due to residents failing to heed evacuation advisories. Risk communication research suggests that flood warnings need to be more vivid, contextualized, and visualizable, in order to engage the message recipient. This paper makes the case for the development of a low-cost augmented reality tool that enables individuals to visualize, at close range and in three-dimension, their homes, schools, and places of work and worship subjected to flooding (modeled upon a series of federally expected flood hazard levels). This paper also introduces initial tool development in this area and the related data input stream.

1. INTRODUCTION

Annually thousands of people fail to evacuate from water hazard events. As will be discussed, a number of cognitive models support the proposition that augmented/virtual reality (AR/VR) could help overcome persistent cognitive roadblocks to effective risk communication and subsequent evacuation actions. Virtual reality enables the user to explore a virtual representation of a specific space (real or synthetic). Augmented reality layers simulated features on top of actual scenes. In this paper, these sets of technologies will be treated as one under the term AR/VR. Thus, this paper introduces the related literature and initial efforts in related tool development in AR/VR for communicating, and preparing for, the risks of extreme weather events such as tropical cyclones, floods, and storm surges.

2. CONCEPTUAL JUSTIFICATIONS FOR THE USE OF AR/VR

Risk preparedness is a factor of what scholars sometimes refer to as experiential processing (i.e., the extent to which previous events have primed the public to understand the nature of risk (Marx et al., 2007). This is directly related to an individual's or group's memories of past events (Vasileiadou and Botzen, 2014) and, less directly, related to institutional memory, which pertains to how events, even those in the distant past, are reflected in customs or local institutions that have become embedded in a place (Sharma et al., 2013). Problems arise when the risks are of such nature or magnitude that the populace has never experienced anything like it. There is mounting evidence that climate change is creating risk scenarios of just this sort, as climate models suggest that extreme weather events (e.g., tropical cyclones) will increase in frequency and severity in the coming decades (Emanuel, 2013; Grinsted et al., 2013). Thus, the challenge is that people will not have a good idea of what such an event would be like, but also that the prospect of such an event happening to them will seem less likely.

As AR/VR can be thought of as a means for creating the sensation that the user is somewhere where they are not (Virtual Reality Society, 2017), such technology could play a role in making an event easier to imagine and picture is a compelling prospect. With AR/VR visualization, a disaster event may also become more realistic, especially if the technology supports vivid, detailed imagery. For example, Haynes, Hehl-Lange, and Lange (2018), developed a mobile augmented reality app that allows one to select a setting and, using a smartphone, simulate a pre-specified flood onto the scene. Lieske (2012) discusses how such visualization tools make up for, or complement, one's understanding of the severity of a risk. However, it is an open question whether such visualization aids motivate individuals to act, since the person's personal threshold for action has to be met, along with a perception that one's actions can be effective (Lieske, 2012).

These notions are corroborated by various cognitive models, such as the elaboration–likelihood model (ELM) [Petty and Cacioppo 1984, 1986]. ELM proposes a theory about processes by which a person receives a message from another and how and why such a message is internalized by the person and influences behavior. The likelihood that a person can plan for an event in advance is increased to the extent that the risk information is perceived as self-relevant and the consequences of the event perceived as threatening. AR/VR can, conceptually, aid in this respect, especially if an event like a flood is visualized at a person's specific locale (increasing self-relevance) and makes the magnitude of a large event more realistic (increasing consequentiality).

AR/VR may enhance risk communication in yet another way, which is to increase a person's perception of both the ability to act and the efficacy of any potential risk-reductive action, as proposed by the Theory of Planned Behavior (Ajzen, 1991). Conceivably, viewing a realistic rendering of one's own home under flood conditions locale may allow the person to more explicitly imagine how to react in such a situation, which can likewise increase the perceived efficacy of such action. By increasing the degree of realism of one's mental model or imagery of the hazard scenario, the consequences of the event, and any risk-reductive action might likewise be more tangible.

Lastly, the Relational Theory of Risk Communication suggests that AR/VR might help in transforming community residents from passive recipients of technical information to active agents who are more willing to act upon their situation (Lejano et al., 2018; Lejano et al., 2020). Again, the qualities of the AR/VR interface could be an important element. If the user could move around a simulated environment in different ways, it may also increase the person's sense of empowerment.

This suggests that the quality (i.e., the degree of resolution and detail) of the simulated image may be key to enhancing the above effects. The simulations should also be of actual places – ideally, near the user's home or place of work. The ability to toggle or move through the scene (simulating actual movement) may enhance the tangible aspects of the simulation.

Research in environmental psychology routinely demonstrates that people's perception of their environment fosters their ability to behave with improved cognition and stronger emotion. These stronger connections to their environment can also change the way they behave and interact in social environments (Russel et al., 1982). With AR/VR, the user can trick the mind, immerse the senses in a different way, and provide context to a place without the need of being physically present. This utility is being explored in many areas of science, including ecology, urban systems, environment, and climate. To create realistic and immersive experiences, several components are required. Stereoscopic displays are a preliminary and basic component and simply provide a 3D visualizer for the user, typically through a head mounted display. In order to replicate the sensation of moving one's body and observe the surroundings, some sort of motion tracking hardware is required. Something like a gyroscope or accelerometer allows the software to pick up on the user's body motion. Input devices, such as joysticks or keyboards, can let the user navigate the virtual space through gestures and movement. Lastly a visual platform is required to run the application (Velev and Zlateva, 2017).

The literature is still sparse regarding how qualities of the AR/VR images and interfaces affect the effectiveness of the simulation and through what cognitive pathways these interfaces work to enhance risk perception and readiness-to-act.

3. BACKGROUND: THE EVOLUTION OF AR/VR

AR/VR is one of the many modern technological breakthroughs that has surged in popularity over the past decade. Although there are many applications for this technology, the most commonly known is its use in conjunction with video games. Consoles like the Sony Playstation created early iterations of consumer grade virtual reality headsets, such as the company's namesake VR headset released in 2016, selling over 4.2 million units globally as of March 2019 (Shuman, 2019). Like other gaming studios, Sony entered the consumer level VR market following the early success of Oculus. Oculus VR raised nearly $2.5 million dollars through a Kickstarter crowdfunding campaign in 2012 for its first generation device the Oculus Rift. In March 2014, Facebook purchased Oculus for $2 billion (Orland, 2014). While the 2010s marked the heyday for AR/VR, attempts at virtual reality can be traced back as early as the 1830s. In 1838, scientists such as Charles Wheatstone and David Brewster contributed to the development of a stereoscope. The original design and construction of this device was later used to make the View-Master stereoscope, where users could look at images through a viewfinder. Subsequently, Google used this design to make Google Cardboard, a low budget system that has a smartphone mount, providing low cost VR accessibility to users (Virtual Reality Society, 2017).

3.1 Applications of AR/VR to Emergency Preparedness

In 2005, Beckett and Shaffer explored the use of gaming environments (in their case SimsWorld) and its effects on informed city planning. Their paper argues that the use of virtual models, through video games and augmented reality, can provide students with informed representations for problem solving in complex environments. The idea of virtual reality providing experiential problem-solving activities is actively being explored in disaster preparedness and emergency response. For example, Schuurink and Toet (2010) employed virtual models by using the game Second Life, conceptually similar to SimsWorld, to have participants navigate evacuation routes in disaster situations. Although their experimental group was small (n=22), their study found that the participant's viewing perspective aided their search and evacuation missions within the module.

Bernardes et al. (2015) explained how AR/VR can train emergency workers (and the public) by simulating the three phases of an emergency situation: risk perception (i.e., recognizing contextual cues for action), movement/behavior (i.e., carrying out risk-avoidance or mitigating actions), and post-movement (i.e., steps proceeding after initial evacuation or other response).

Beyond the applications of random study participants, AR/VR is being used to train medical personnel, from everyday tasks to mass casualty events, and can provide a safe learning and practice space for personnel in this field. Farra et al. (2015) explored AR/VR and its role in situated cognition. When the learner enters an augmented reality, they experience embeddedness, extension, and embodiment (Farra et al., 2015). With practice and repetition, the learner will have an increased understanding of the simulation, hopefully translating to better performance in a real disaster scenario. Farra et al. (2015) used AR/VR to have nursing students undergo disaster training, with half of their group using web-based modules instead. The participants found the AR/VR experience to be more engaging and provided a realistic method of practice as opposed to watching informational PowerPoints and videos via web-based training.

Andreatta et al. (2010) explored the differences in efficacy between AR/VR and standardized patient drills when providing triage during mass casualty incidents. While they did not find significant differences between the two, AR/VR seemed to be favored, as it was cost-effective compared to using live volunteers in disaster drills. The feasibility and repeatability of AR/VR makes it a realistic alternative to live drills and may lead to better results in live drills when used in tandem. Technologies like this aim to ameliorate the negative impacts that cognitive dissonance can have during mass casualty events. This dissonance can result from sensational overload and is likely to negatively impact honed skill sets in emergency medical personnel (Andreatta et al., 2010). Applied contexts such as AR/VR serve as better alternatives to non-applied contexts, such as web-based or lecture formatted training that

This contribution has been peer-reviewed.
https://doi.org/10.5194/isprs-archives-XLIV-M-2-2020-37-2020 | © Authors 2020. CC BY 4.0 License.

has been explored in other research in emergency preparedness. AR/VR is growing in popularity in the educational sphere in a multitude of disciplines. In another case, AR/VR was used to train obstetrics teams, resulting in significantly less errors in real time applications (Merién et al., 2011).

In the case of the United States and disaster and emergency preparedness, several government agencies are taking advantage of AR/VR technology training. For example, the New York City Office of Emergency Management and Los Angeles Police Department are actively using simulation systems to train staff on emergency operation, navigating urban environments, and communication with colleagues during these events (Hsu et al., 2013). Recent terrorist events and natural disasters in the past few decades have prompted municipalities to improve their staff's preparedness, and AR/VR technology provides the potential for an efficient and cost sensible training method. In the case of natural disasters and their increased occurrences due to climate crises, there is a growing demand for efficacious AR/VR training programs that will help not just first responders, but citizens who have a rising susceptibility to natural disasters as the climate crisis worsens.

3.2 Application of VR to Risk Communication

Although one cannot replace first-hand experiences of natural hazards, it is thought that, as the simulation is credible, AR/VR may prime the public to better perceive risks and plan for them (Mitchell and Cutter, 1997; Basic et al., 2003; Maidl and Buchecker, 2014; Havenith et al., 2019)[1].

Best practices of crisis and risk communication rely on initial notifications commencing prior to the event. The message's specificity, clarity, and certainty determine how effective it is (Markwart et al., 2019). It is then up to the citizen to make a rational decision based on the magnitude of risk associated with the event. Modes of risk communication can vary from generalized, regional warnings (e.g., "floods ranging from 1 to 3 feet expected on the east coast") to context-specific ones ("2 foot storm surge along seafront between 5th St. and 8th St."). Risk messages also vary depending on mode of expression, the simplest being plain text (i.e., words on paper or on a screen). Often, text will be supplemented with graphic aids such as maps which give two-dimensional place-specific hazard information. Flood and storm surge maps have been shown to aid the public's understanding of the risk, though people display differing degrees of comfort/skill at interpreting maps (Morrow et al., 2015). In this article, we explore the possibilities of 3D flood representations for improving viewer comprehension even further.

However, if the individual has no prior experience with certain types of disasters or environmental hazards, there is no experience from which to draw. AR/VR training aims to provide experiences to draw from that work in tandem with warning messages, hopefully leading to rational decision making and increased safety of communities. Markwart et al. (2019) tested the efficacy of warning messages through AR/VR by putting participants in virtual storms and having them navigate to shelters following the message being received. All but 1 of their 31 participants made it to a shelter within a safe amount of time. Moreover, participants who received the warning through the app found an escape route more quickly than the control group (Markwart et al., 2019). Though receiving warning messages during storm events is a rudimentary aspect of overall risk communication and best safety practices, this research successfully illustrates the importance of comprehensive communication lines between public authorities and the community in order to ensure safety.
Aside from the utility of AR/VR as a medium for training personnel and community members, virtual environments are useful for observing communities and understanding their susceptibility and need for future adaptations within their social environment or infrastructure. An example of using this technology to generate input to government planning risk management activities to mitigate damage was done by Avagyan et al. (2018) for the Ministry of Emergency Situations of Armenia. That study was undertaken to evaluate the relevant physical, social, and economic components in Armenia to better understand the population's coping and adaptive capacity, as well as its general vulnerability. Armenia is prone to earthquakes, landslides, and an array of other hazardous natural processes such as strong wind and drought. Using a simulated environment, spatial evaluations were conducted to collect data on hazard assessment for various environmental risks. A vulnerability index was then applied to each region, as a composite of physical, social, and economic risk and certain communities were targeted as being especially vulnerable. Similarly, Wang et al. (2019) simulated the flooding levels of a Chinese city with and without additional infrastructure.

The research team is planning on translating these visualizations to AR/VR headsets to share with stakeholders and community members to encourage the development of crucial infrastructure improvements by showing how such investments can mitigate flooding (Wang et al., 2019).

With a successful track record in the training of equipment operators and medical personnel, AR/VR would appear to hold the potential to ameliorate some of the impacts of natural disasters and the burden placed on first responders by better educating communities and providing community members safe immersions into disaster scenarios, allowing direct perception of these environments more effectively than other methods (Hu et al., 2018).

Studies in the mid to late 2000s, such as Beckett and Shaffer (2005) and Schuurink and Toet (2010) utilized GIS or even video games, such as Sims or Second Life, to test risk assessment workshops. These studies were useful and found success in teaching participants about risk preparedness, how to navigate escape routes or hazardous obstacles during disaster scenarios, and visualize the impact of natural disasters. Arguably AR/VR of impacted areas could provide localized context to these events and influence behavior and attitudes towards disaster preparedness. Older versions of flood maps, including physical mediums and GIS displays, are not readily understandable to the general public, who generally have difficulty deriving meaning from them (Macchione et al., 2019; Hagemeier-Klose and Wagner, 2009). To the extent that a 3D visualization more closely resembles the real situation, it has the possibility of overcoming limitations posed by 2D maps to those for whom the latter are too abstract or non-realistic (Lieske, 2012).

Changing perceptions can serve as a keystone solution for shifting the attention of the public eye. Solinkska-Nowak et al. (2018) compiled an overview on risk management through virtual reality games which provide training for different scenarios. One visualization called *VR for a New Climate* had

[1] https://games4sustainability.org/gamepedia/forest-at-risk/

https://doi.org/10.5194/isprs-archives-XLIV-M-2-2020-37-2020 | © Authors 2020. CC BY 4.0 License.

participants operate a variety of devices and perform functions during a flood disaster scenario, spanning from activating alarm bells or loading relief kits for transport. Other examples include *Forest@Risk* which emphasizes the idea of the "tragedy of the commons" and the impacts that the climate crisis will have on natural resources. Applications of this game in workshops showed interesting results as the game provides communication lines between players which leads to complex decision making and cooperation in order to conserve natural resources for a community. The social dimension that activity-based games or interactive AR/VR workshops possess are one of the utility's most potent advantages when creating or improving the way communities handle natural disasters and build on their resiliency. Not only do these activities provide tangible lessons within themselves, but a lot of learning happens after the fact during debriefing sessions (Crookall 2010) where participants are able to discuss the pros and cons of the workshop or scenario within the game model, allowing them to repeat the experiment and further build on this social component and its relation to risk communication and preparedness.

Evans et al. (2014) found efficacy in their experiment when they presented 3D visualizations to the community of Exeter, England during a 50th anniversary commemoration of historic flooding in 1960. Participants in their experiment described the visualizations as a wake-up call and were reminded that though modern infrastructure may provide some protection, increasing intensity of rainfall and storm surge due to climate crises may make this problem closer to home than previously anticipated. One notable point from the participants was that the realization of flood damage to local landmarks was "chilling and sobering". Responses like these are easy to derive from experiments that utilize VR technology since the visualizations are easy to interpret compared to older versions of 3D modeling and enhanced mapping, especially when technology such as drones and LIDAR data are utilized to construct highly detailed replicas of familiar locales. Two years after Evans et al. (2014) conducted their research, high-flow flooding occurred in the Exeter region, and the UK's Environment Agency reported heightened awareness of protocol, which the authors attributed to flood visualization workshops.

Macchione et al. (2019) used a similar approach in Cosenza, Italy which has been prone to flooding as early as the 1590s and as recent as the 1950s. This time span provided an opportunity for the researchers as they were able to better understand the perception of participants who had experienced flooding in the 1950s, and the younger generations who have not experienced any. Through vivid visuals, Macchione et al. (2019) were able to contextualize the risk and substituted the visuals for the experience of flooding that older generations had. A portion of this workshop focused on emergency preparation and participants were able to focus on risk communication within the community, such as when certain roads were impassable and climbing to rooftops of buildings for safety located throughout the city.

The results from experiments like these suggest that effective and impactful virtual reality programs should be tailored to local areas to provide a social component to the participants, especially if the given area is prone to flood risk and storm surge. Enhanced virtual and augmented reality technologies provide researchers with opportunities to create immersive environments that can lessen the disconnect that many citizens have from the consequences of the climate crisis, especially those who have not had experience with natural disasters and are now facing potential scenarios due to worsening climate conditions. Aside from community members, policy makers and public stakeholders are benefiting from the use of VR technology. An example of the initial development of such a tool is next explained.

4. TOOL INTRODUCTION

4.1 Overview statement of the goals of the technology

The primary goal of the UrbanARK system was to create an inexpensive means to simulate 3D environments with flood inundation data to better contextualize what particular flooding models are trying to convey. Using commercially off-the-shelf (COTS) hardware and open source packages increases the likelihood that this would be accessible to more communities than development of an entirely new piece of hardware. As such the UrbanARK tool has the user wear a cheap readily available/easy to build AR/VR viewer for a smart mobile phone device. The smart phone renders the 3D environment for the user, projected through a set of lenses, which creates the 3D effect. To be operational the smart phone must be equipped with a set of commonly incorporated sensors. These include an accelerometer, a gyroscope, and a magnetometer. Together these convey user location within the space and head orientation to simulate the heads movements through the 3D virtual environment. The work file is shown in Figure 1.

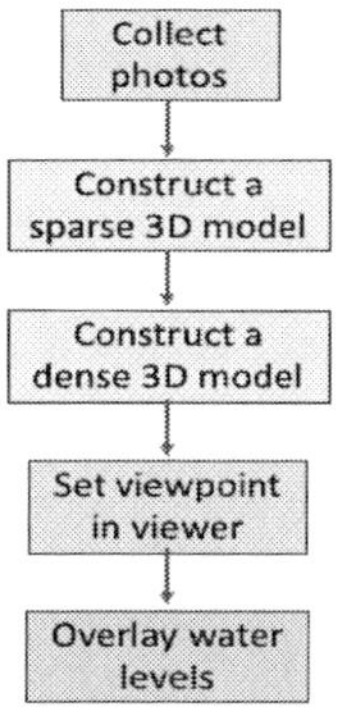

Figure 1. Workflow Overview

4.2 Input data

The input data consist of a set of images of a point of interest captured using a COTS camera digital single lens reflex (DSLR) or mobile phone camera, or an inexpensive Unmanned Aerial Vehicle with an inbuilt or custom mounted camera. Using a device enabled with a global positioning system (GPS) helps scale the environment automatically, so that buildings, roads, and other easily recognizable features are at a real-world scale during the creation of the 3D model. Preferably, the GPS data are contained within the Exchangeable Image File format (EXIF) meta-data of the image file itself, as is readily available in many smartphones. This is helpful, especially for larger 3D model renders with limited computing power, as the images can be clustered and processed on any available hardware, and then easily joined with other clusters as the clusters are geo-rectified to a specific GPS scale space (e.g. WGS84 or UTM in meters).

4.3 Model Generation and Software

Model generation is handled by third party software packages. To allow any community to build a model for viewing, the

https://doi.org/10.5194/isprs-archives-XLIV-M-2-2020-37-2020

openMVG (Moulon et al., 2016) was selected. This software framework is fully open source and contains software plugins to help data generated from other 3D reconstruction or densification packages. openMVG (Moulon et al., 2016) can be used with simple Python language scripting in the command terminal or Regard3D (Regard3D, 2020), a graphical user interface being built to support openMVG (Moulon et al., 2016). The model generation works by first organizing the images to be used in the reconstruction by putting them into a folder. The openMVG_main_SfMInit_ImageListing then looks at the image directory and creates the initial openMVG structure for the framework. This is followed by image feature extraction (identifying interesting features within the images) using openMVG_main_ComputeFeatures. If there are a large number of images, images are clustered based on GPS position data using openMVG_main_ListMatchingPairs. After this there is the image feature descriptor matching stage using openMVG_main_ComputeMatches. This stage requires some experimentation by the user, to set the matching ratio. If the matching ratio is too high, sufficient points may not exist to do the reconstruction. If the matching ratio is too low, there may be a lot of noise. Once the matching is complete, then openMVG_main_GlobalSfM is used to generate the sparse 3D point cloud. This sparse 3D model shows whether the image descriptor matching ratio was appropriate. If the model is usable, the model is then prepared for densification using the Multiple View Environment (MVE) (Fuhrmann et al., 2014) package.

Densification is done by first using the framework converter openMVG_main_openMVG2MVE2 in the openMVG package. This generates the file structure for (Fuhrmann et al., 2014). The next step is to start the densification using the MVE (Fuhrmann et al., 2014) dmrecon program, which estimates depth (distance from camera) for each pixel in each image. The last step is the dense map merging using MVE's scene2pset program. This takes all the dense depth maps and uses them to project the pixels from the images into a 3D scale space. The output from this is a dense point cloud generally in the PLY format (Turk, 1994).

Now that a 3D point cloud model has been generated, the 3D model viewer will need to be generated. For the 3D rendering Potree (Schuetz et al., 2020a) and Three.js (Cabello et al., 2020) are used in the browser to render the 3D model and flood inundation information in AR/VR. The VR effect is created by using the Three.js StereoEffect renderer. PotreeConverter (Schuetz et al., 2020b) is used to convert the data from its point cloud format into the Potree native Octree format. By doing this, Potree can generate the points that should be viewed into a Level Of Detail pyramid (LOD pyramid) that optimizes the rendering by only showing close points in full detail, and rendering fewer points at locations that are further away. This allows large models to be shown without the need to have expensive hardware to view them.

Now that the 3D environment can be viewed, the next step is to add the flood model data (Fig. 2). The first step is setting a reasonable height of the AR/VR viewers viewpoint. For this, a simple statistical average height of a person in the United States was used (161.5cm for females + 175.3cm for males = 168.4cm or 5 feet and 6.2992 inches). The height is then set in Potree by taking the height of the point cloud floor where the viewpoint is and adding 168.4cm to it (the model scale space is in metric). Future applications would allow for different points of view and perspectives specific to individual users (e.g. children, people with visual impairments).

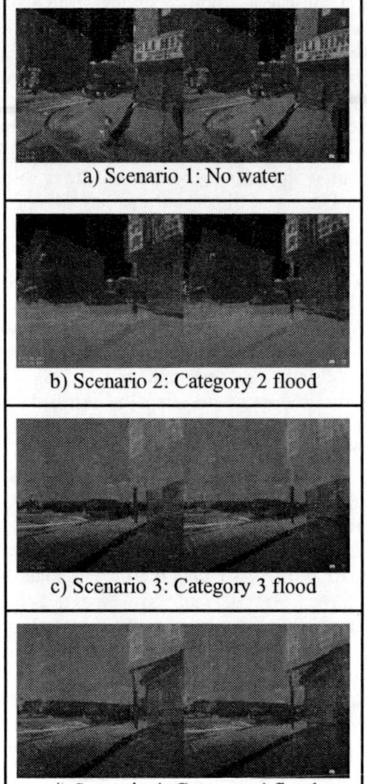

a) Scenario 1: No water

b) Scenario 2: Category 2 flood

c) Scenario 3: Category 3 flood

d) Scenario 4: Category 4 flood

Figure 2. Viewer flood levels

The last step is setting the flood level within the model. This was generated by looking at the Federal Emergency Management Agency (FEMA, 2020) flooding maps. The spatial extent likely to be covered by a specific category of flooding is mapped in the 3D space. This space is then occupied by a polygon (Turk 1994). The flooding characteristic is created by mapping the category level to the z-height value of the polygon. If the entire scene is affected, then a cube is mapped across the entirety of the scene and then the same height designator is used. The height of the polygon or cube is set to that of the ground level of the point cloud.

Next, a timer is created in JavaScript that modifies the height of the polygon or cube to increase/rise to a particular FEMA (FEMA, 2020) map flood level (Figure 2). A texture can be applied to the polygon/cube for water, or a simple colour can be used to make the polygon or cube blue for the water level. The prototype system was built using 4 different scenarios. First is the 3D model with no flooding. The second scenario is the 3D model with category 2 flood levels. The third scenario is the 3D model with category 3 flood levels. The final scenario is the 3D model with category 4 flood levels. Once the final scenario has been shown, the system then loops back around to the first scenario. By using Three.js (Cabello et al., 2020), the user can press w,s,a,d on the keyboard to move forward, back, left and right in the 3D environment.

This contribution has been peer-reviewed.
https://doi.org/10.5194/isprs-archives-XLIV-M-2-2020-37-2020

4.4 Capabilities

To promote wide adoption of UrbanARK as a flood risk communication tool, the entire system is open source, and the point cloud can be scaled to suit the processing power of the smartphone device being used. The system is compatible with higher end AR/VR equipment such as Oculus (Oculus, 2020) or HTC (HTC VIVE. 2020) headsets but is designed more for mid-range and better smartphones. The AR/VR headset enables a full, 360-degree 3D AR/VR environment (Figure 3).

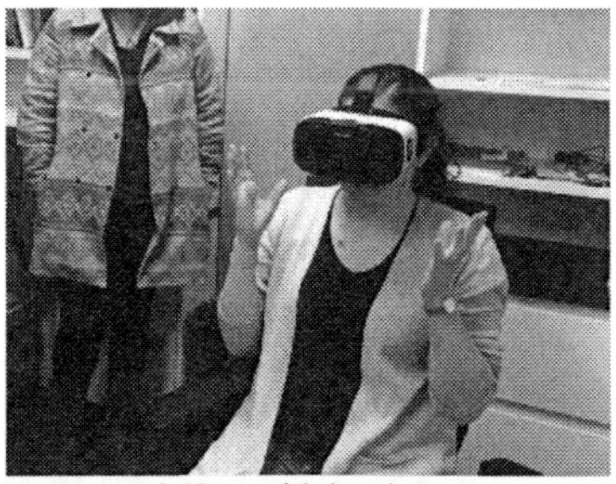

a) User with headset on

b) Scene visible within viewer

Figure 3. Viewer in use.

5. CONCLUSIONS

This system is compatible with any web browser that is webGL (Khronos Group Inc., 2020) 1.0 compatible. Future envisioned capabilities will allow users to walk around the 3D VR environment. Such additional features could permit users to test evacuation routes. While the alpha version of this AR/VR tool is ready for testing, the current COVID-19 situation prevents that. In the interim, the development team has been working on an online version.

There has been increasing use of AR/VR and 3D visualization tools for risk communication and emergency preparedness. However, deeper research on this is lacking. To date, little work has been done as to what features or qualities of the visualization tool are more effective in hazard risk communication. Specifically, there is inadequate work on what cognitive pathways (increased consequentiality, enhanced vividness, affordance of individual agency, etc.) are triggered by use of these tools. There is also need for research on collaborative forums (e.g., PAR, scenario building, etc.) for community engagement. The low-cost UrbanARK AR/VR flood simulation tool may serve as a means to understanding some of these deeper issues.

ACKNOWLEDGEMENTS

Funding provided by the U.S. National Science Foundation (Award 1826134), Northern Ireland Trust (Award R3078NBE), and Science Foundation Ireland (Award 17/US/3450).

REFERENCES

Andreatta, P. B., Maslowski, E., Petty, S., Shim, W., Marsh, M., Hall, T., Stern, S., Frankel, J., 2010: Virtual Reality Triage Training Provides a Viable Solution for Disaster-preparedness. *Academic Emergency Medicine,* 17(8), 870-876. doi.org/10.1111/j.1553-2712.2010.00728.x.

Avagyan, A., Manandyan, H., Arakelyan, A., Piloyan, A., 2018: Toward a disaster risk assessment and mapping in the virtual geographic environment of Armenia. *Natural Hazards,* 92(1), 283-309. doi.org/10.1007/s11069-018-3208-0.

Ajzen, I., 1991: The theory of planned behavior. *Organizational Behavior and Human Decision Processes*, 50(2), 179-211. doi.org/10.1016/0749-5978(91)90020-T.

Beckett, K. L., Shaffer, D. W., 2005: Augmented by Reality: The pedagogical praxis of urban planning as a pathway to ecological thinking. *Journal of Educational Computing Research*, 33(1), 31-52. doi.org/10.2190/d5yq-mmw6-v0fr-rnjq.

Basic, F., Cartwright, W., Handmer, J., 2003: Geographic visualization tools for communicating the risk of floods. *In Proceedings of the 21st International Cartographic Conference*, 2033-2044. Durban, South Africa: International Cartographic Association.

Bernardes, S. M. F., Rebelo, F., Vilar, E., Noriega, P., Borges, T., 2015: Methodological approaches for use virtual reality to develop emergency evacuation simulations for training, in emergency situations. *Procedia Manufacturing*, 3, 6313-6320. doi.org/10.1016/j.promfg.2015.07.946.

Cabello, R., and others, 2020: three.js. GitHub. https://github.com/mrdoob/three.js/.

Crookall, D., 2010: Serious games, debriefing, and simulation/gaming as a discipline. *Simulation and Gaming, 41*(6), 898-920. doi.org/10.1177/1046878110390784.

Emanuel, K. A., 2013: Downscaling CMIP5 climate models shows increased tropical cyclone activity over the 21st century. *Proceedings of the National Academy of Sciences of the United States of America*, 110(30), 12219–12224. doi.org/10.1073/pnas.1301293110.

Evans, S. Y., Todd, M., Baines, I., Hunt, T., Morrison, G., 2014: Communicating flood risk through three-dimensional visualization. *Proceedings of the Institution of Civil Engineers – Civil Engineering,* 167(5), 48-55. doi.org/10.1680/cien.13.00020.

Hagemeir-Klose, M., Wagner, K., 2009: Evaluation of flood hazard maps in print and web mapping services as information tools in flood risk communication. *Natural Hazard and Earth System Sciences,* 9(2), 563-574. doi.org/10.5194/nhess-9-563-2009.

Farra, S. L., Miller, E. T., Hodgson, E., 2015: Virtual reality disaster training: Translation to practice. *Nurse Education in Practice,* 15(1), 53-57. doi.org/10.1016/j.nepr.2013.08.017.

Federal Emergency Management Agency, 2020. Home | FEMA.gov. https://www.fema.gov/

Fuhrmann, S., Langguth, F., Goesele, M., 2014: *In Proceedings of the Eurographics Workshop on Graphics and Cultural Heritage*, 11-18. Darmstadt, Germany: Eurographics Association.

Grinsted, A., J. C. Moore, and Jevrejeva, S., 2013: Projected Atlantic hurricane surge threat from rising temperatures. *Proceedings of the National Academy of Sciences of the United States of America.,* 110(14), 5369–5373. doi.org/10.1073/pnas.1209980110.

Havenith, H.-B., Cerfontaine, P., Mreyen, A.-S., 2019: How virtual reality can help visualise and assess geohazards, *International Journal of Digital Earth*, 12(2), 173-189. doi.org/10.1080/17538947.2017.1365960.

Haynes, P., Hehl-Lange, S., & Lange, E., 2018. Mobile augmented reality for flood visualisation. Environmental modelling & software, 109, 380-389.

Hsu, E. B., Li, Y., Bayram, J. D., Levinson, D., Yang, S., Monahan, C., 2013: State of Virtual Reality Based Disaster Preparedness and Response Training. *PLOS Currents, 1.* doi.org/10.1371/currents.dis.1ea2b2e71237d5337fa53982a38b2aff.

HTC VIVE, 2020: VIVE™ | Discover Virtual Reality Beyond Imagination. https://www.vive.com/eu/.

Hu, Y., Zhu, J., Li, W., Zhang, Y., Zhu, Q., Qi, H., . . . Zhang, P., 2018: Construction and Optimization of Three-Dimensional Disaster Scenes within Mobile Virtual Reality. *ISPRS International Journal of Geo-Information,* 7(6), 215. doi.org/10.3390/ijgi7060215.

Khronos Group Inc., 2020: Khronos royalty-free open standards for 3D graphics, Virtual and Augmented Reality, Parallel Computing, Machine Learning, and Vision Processing. WebGL. https://www.khronos.org/.

Lejano, R. P., Casas Jr, E. V., Montes, R. B., Lengwa, L. P., 2018: Weather, climate, and narrative: a relational model for democratizing risk communication. *Weather, climate, and society*, 10(3), 579-594. doi.org/10.1175/WCAS-D-17-0050.1.

Lejano, R. P., Rahman, M. S., Kabir, L., 2020: Risk Communication for Empowerment: Interventions in a Rohingya Refugee Settlement. *Risk Analysis*. doi.org/10.1111/risa.13541.
Lieske, D. J., 2012: Towards a framework for designing spatial and non-spatial visualizations for communicating climate change risks. *Geomatica*, 66(1), 255-265. doi.org/10.5623/cig2012-006.

Macchione, F., Costabile, P., Costanzo, C., Santis, R. D., 2019: Moving to 3-D flood hazard maps for enhancing risk communication. *Environmental Modelling & Software,* 111, 510-522. doi.org/10.1016/j.envsoft.2018.11.005.

Maidl, E., Buchecker, M., 2014: Raising risk preparedness through flood risk communication. *Natural Hazards and Earth System Sciences Discussions*, 15(7), 1577-1595. doi.org/10.5194/nhess-15-1577-2015.

Markwart, H., Vitera, J., Lemanski, S., Kietzmann, D., Brasch, M., Schmidt, S., 2019: Warning messages to modify safety behavior during crisis situations: A virtual reality study. *International Journal of Disaster Risk Reduction,*38, 101235. doi.org/10.1016/j.ijdrr.2019.101235.

Marx, S. M., Weber, E. U., Orlove, B. S., Leiserowitz, A., Krantz, D.H., Roncoli, C., Phillips, J., 2007: Communication and mental processes: Experiential and analytic processing of uncertain climate information. *Global Environmental Change*, 17(1), 47–58. doi.org/10.1016/j.gloenvcha.2006.10.004.

Merién, A.E., van de Ven, J., Mol, B.W., Houterman, S., Oei, S.G., 2011: Multidisciplinary Team Training in a Simulation Setting for Acute Obstetric Emergencies. *Obstetric Anesthesia Digest,* 31(2), 83-84. doi.org10.1097/01.aoa.0000397113.95428.cb.

Mitchell, J.T., Cutter, S.L., 1997: *Global Change and Environmental Hazards: Is the World Becoming More Disastrous? Hands-On! Developing Active Learning Modules on the Human Dimensions of Global Change.* Association of American Geographers, Washington, DC, USA.

Morrow, B. H., Lazo, J. K., Rhome, J., Feyen, J., 2015: Improving storm surge risk communication: Stakeholder perspectives. *Bulletin of the American Meteorological Society*, 96(1), 35-48. doi.org/10.1175/BAMS-D-13-00197.1.

Moulon, P., Monasse, P., Marlet, R., and others, 2016: openMVG. GitHub. https://github.com/openMVG/openMVG.

Oculus, 2020: Oculus | VR Headsets & Equipment. https://www.oculus.com/.

Orland, K., Ars Technica, 2014. Facebook purchases VR headset maker Oculus for $2 billion. https://arstechnica.com/gaming/2014/03/facebook-purchases-vr-headset-maker-oculus-for-2-billion/ (25 March 2014).

Petty, R. E., Cacioppo, J.T., 1984: Source factors and the elaboration likelihood model of persuasion. *Advances in Consumer Research*, 11, 668–672. https://www.acrwebsite.org/volumes/6328/volumes/v11/NA-11.

Petty, R.E., Cacioppo, J.T., 1986: *Communication and Persuasion: Central and Peripheral Routes to Attitude Change.* Springer-Verlag New York.

Regard3D, 2020: Regard3D, a free and open source structure-from-motion-program. Version 1.0.0.https://www.regard3d.org/ (14 March 2019).

Rossol, N., Cheng, I., Bischof, W. F., Basu, A., 2011: A framework for adaptive training and games in virtual reality rehabilitation environments. *Proceedings of the 10th International Conference on Virtual Reality Continuum and Its Applications in Industry - VRCAI 11.* doi.org/10.1145/2087756.2087810.

Russell, J. A., Ward, L. M., Pratt, G. 1982: Affective Quality Attributed to Environments: A Factor Analytic

Study. *Environment and Behavior,* 13(3), 259-288. doi.org/10.1177/0013916581133001.

Schuetz, M., and others, 2020a: potree. GitHub. https://github.com/potree/potree.

Schuetz, M., and others, 2020b: PotreeConverter. Github. https://github.com/potree/PotreeConverter.

Shuman, S., Playstation Blog, 2019: PlayStation VR: The Next Wave of Games Coming in Spring and Summer 2019. blog.us.playstation.com/2019/03/25/playstation-vr-the-next-wave-of-games-coming-in-spring-and-summer-2019/ (26 March 2019).

Schuurink, E. L., Toet, A., 2010: Effects of Third Person Perspective on Affective Appraisal and Engagement: Findings from SECOND LIFE. *Simulation & Gaming,* 41(5), 724-742. doi.org/10.1177/1046878110365515.

Sharma, U., Patwardhan, A., Patt, A.G., 2013: Education as a determinant of response to cyclone warnings: Evidence from coastal zones in India. *Ecology and Society*, 18(2), 18. doi.org/10.5751/ES-05439-180218.

Solinska-Nowak, A., Magnuszewski, P., Curl, M., French, A., Keating, A., Mochizuki, J., . . . Jarzabek, L., 2018: An overview of serious games for disaster risk management – Prospects and limitations for informing actions to arrest increasing risk. *International Journal of Disaster Risk Reduction, 31*, 1013-1029. doi.org/10.1016/j.ijdrr.2018.09.001.

Turk, G., 1994: Polygon File Format (PLY) Family. Digital Preservation at the Library of Congress. https://www.loc.gov/preservation/digital/formats/fdd/fdd000501.shtml (2019).

Vasileiadou, E., Botzen, W.J.W., 2014: Communicating adaptation with emotions: The role of intense experiences in raising concern about extreme weather. *Ecology and Society,* 19(2), 36. doi.org/10.5751/ES-06474-190236.

Velev, D., Zlateva, P., 2017: Virtual Reality Challenges in Education and Training. *International Journal of Learning and Teaching*, 3(1). doi.org/10.18178/ijlt.3.1.33-37.

Virtual Reality Society, 2017. History Of Virtual Reality. https://www.vrs.org.uk/virtual-reality/history.html (26 March 2019).

Wang, C., Hou, J., Miller, D., Brown, I., Jiang, Y., 2019: Flood risk management in sponge cities: The role of integrated simulation and 3D visualization. *International Journal of Disaster Risk Reduction,* 39, 101139. doi.org/10.1016/j.ijdrr.2019.101139.

COMPARISON OF IMAGE ENHANCEMENT TECHNIQUES FOR RAPID PROCESSING OF POST FLOOD IMAGES

M. Harichandana [1], V. Sowmya [1], V.V. Sajithvariyar [1], Ramesh Sivanpillai [2, *]

[1] Center for Computational Engineering and Networking, Amrita School of Engineering, Amrita Vishwa Vidyapeetham, Coimbatore, TN 641 112, India - mhchandana@gmail.com, (v_sowmya, vv_sajithvariyar)@cb.amrita.edu
[2] Wyoming GIS Center, University of Wyoming, Laramie, WY 82072, USA (ORCID: 0000-0003-3547-9464) – sivan@uwyo.edu

KEY WORDS: Floods, Disaster response, Satellite images, Image contrast, Enhancement algorithms

ABSTRACT:

Satellite images are widely used for assessing the areal extent of flooded areas. However, presence of clouds and shadow limit the utility of these images. Numerous digital algorithms are available for enhancing such images and highlighting areas of interest. These algorithms range from simple to complex, and the time required to process these images also varies considerably. For disaster response, it is important to select an algorithm that can enhance the quality of the images in relatively short time. This study compared the relative performance of five traditional (Histogram Equalization, Local Histogram Equalization, Contrast Limited Adaptive Histogram Equalization, Gamma Correction, and Linear Contrast Stretch) algorithms for enhancing post-flood satellite images. Flood images with different levels of clouds and shadows were enhanced and output generated were evaluated in terms of processing time and quality as measured by Blind/Reference less Image Spatial Quality Evaluator (BRISQUE), a no-reference image quality metric. Findings from this study will provide valuable information to image analysts for selecting a suitable algorithm for rapidly processing post-flood satellite images.

1. INTRODUCTION

Satellite images provide invaluable information on post-disaster conditions to emergency management agencies. Images collected by remote sensors have been effective for evaluating post-disaster conditions (Brivio et al., 2002). Repeat coverage of the affected areas from multiple satellites enable us to track changes over time (Gianinetto et al., 2005), in hard to reach areas at a relatively lower cost, robustness, and with the availability of data preprocessing techniques (Kussul et al., 2008). Satellite-based remote sensing systems offer a bigger potential for assessment of disasters and in their management (Kerle, Oppenheimer, 2002).

Floods are one of the major disasters that impact natural- and built-ecosystems worldwide. Every year flooding events result in the loss of lives, and damages to infrastructure and natural ecosystems. Images collected by active (RADAR) and passive (optical) sensors (Figure 1) on-board satellites are used for mapping and monitoring the extent of floods and changes over time. Optical sensors are limited in terms of collecting data when thick clouds are present. RADAR signals can penetrate cloud cover and collect data on the aerial extent of floods. However, there are relatively more optical sensors than those that collect RADAR data. Hence more optical images are available for monitoring post-flood conditions.

The visual quality of many post-flood optical images can be poor due to factors such as presence of haze, aerosols, and thin clouds and associated shadows (Lee, Lin 2016; Luscombe, Hassan, 1993). Differences in the spectral reflectance from features such as water, forest, vegetation etc. result in poor quality (Singh et al., 2019). Presence of these factors limit the use of optical images for monitoring post-flood conditions.

Figure 1. Landsat 5 Thematic Mapper acquired on 10 May 2011 shows the flooding in the Mississippi River, USA (Image credit: USGS/NASA)

Digital enhancement methods can be used to improve the quality of these optical images to extract useful information from them. Most of these methods focus on manipulating intensity and contrast, de-blurring, de-noising, and edge detection (Bidwai, Tuptewar, 2015). These digitally enhanced

* Corresponding author

This contribution has been peer-reviewed.
https://doi.org/10.5194/isprs-archives-XLIV-M-2-2020-45-2020 | © Authors 2020. CC BY 4.0 License.

images provide additional data to the emergency management agencies for planning rescue and recovery missions.

Numerous image enhancement methods such as Histogram Equalization (Stark, 2000), Linear Contrast Stretching (Gillespie, 1992), Brightness Preserving Bi-Histogram Equalization (Moniruzzaman et al., 2014), Local Histogram Equalization (Kim et al., 1998), Gamma Correction (Huang et al., 2016), Adaptive Histogram Equalization (Pizer et al., 1987), Contrast Limited Adaptive Histogram Equalization (Pizer et al. 1990), Minimum Mean Brightness Error Bi-histogram Equalization (Chen, Ramli, 2003), Dualistic Sub-image Histogram Equalization (Wang et al., 1999) are available for processing low visual quality images. These methods use either spatial or frequency domain for enhancing the quality of the images. To maximize the utility of poor-quality images, emergency management agencies have to enhance those using methods that will produce a quality output in relatively less time. The primary objective of this study was to evaluate the resultant image quality and associated processing time of five frequently used enhancement methods. Post-flood satellite data collected from actual flood events were used to test the performance of Histogram Equalization (HE), Local Histogram Equalization (LHE), Contrast Limited Adaptive Histogram Equalization (CLAHE), Gamma correction (GC), and Linear Contrast Stretch (LCS) methods.

The quality of enhanced images generated by these methods were evaluated using Blind/Reference less Image Spatial Quality Evaluator (BRISQUE) metric. This metric operates in the spatial domain and calculates the loss in naturalness of the image due to distortions by computing locally normalized luminance coefficients using statistics from the enhanced image (Mittal et al., 2012) Results from this study will provide valuable information to emergency management agencies or their partners to select appropriate enhancement methods that will increase the overall quality of the post-flood images in a relatively short amount of processing time.

2. MATERIALS AND METHODS

2.1 Post-flood satellite images

Seven scenes acquired by Landsat 5 Thematic Mapper and Landsat 8 Operational Land Imager were subset to generate 16 multispectral images. These images had poor visual quality due to the presence of thin clouds, haze, and aerosols (Figure 2).

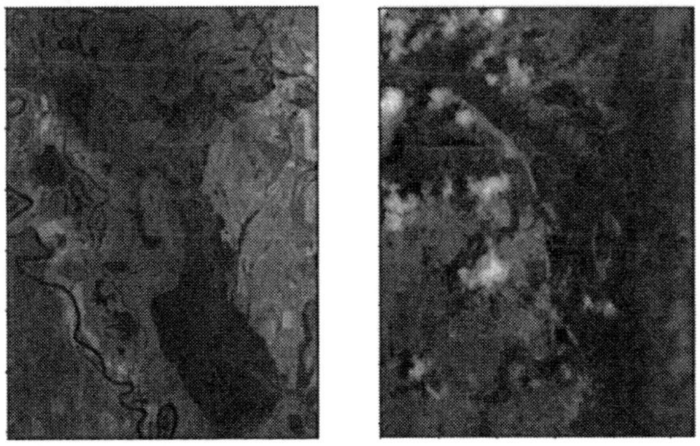

Figure 2. Two of the sixteen post-flood Landsat images used for comparison of the image enhancement algorithms. All images had clouds and haze that reduced their contrast.

First, blue, green, and red bands from these multispectral images were combined to generate 18 true-color images. Next, green and two infrared bands were combined to generate 18 natural or false-color images, resulting in 36 images.

2.2 Image-enhancement methods

Five frequently used image enhancements were selected for this study. One of the criteria used for selecting is their availability in image processing and geospatial software. Brief description of each method is included in the following subsections. All images were processed in Python 3.6 installed in a Windows 10 computer with Intel core i5 (7th generation), 2.71 GHZ, 64-bit processor with 8 GB RAM. Time taken to process each image was recorded and later converted to seconds per megabytes.

2.2.1. Histogram Equalization (HE)

Histogram Equalization (HE) is a spatial domain-based enhancement method that increases the global contrast of an image. For a given input image, HE method automatically determines the transformation function based on the probability density function and maps the grey scale values in the input image to a uniform histogram which is widely spread throughout the available range of radiometric values. For an 8-bit image, the values will be spread from 0 (minimum) to 255 (maximum). Images enhanced by HE can be severe washed out effects due to saturation issues. Also, HE can ignore local details and might not preserve the brightness. Further details on HE method can be found in Stark (2000).

2.2.2. Local Histogram Equalization (LHE)

Local Histogram Equalization (LHE) method uses a sliding window and modifies the values of the central pixel based on the histogram of the subset area. LHE, also known as Adaptive HE, was proposed to overcome one of the limitations of HE method while relies on the histogram for computing the transformation function. The window is moved around the image in both horizontal and vertical direction and in every step the value of the central pixel is modified. A window size has been selected in order to generate an optimal enhanced image. In this study, five window sizes (8x8, 16x16, 32x32, 64x64, and 128x128) were tested. LHE method is computationally intensive and therefore time consuming. Also, LHE can lead to over amplification of noise in the relatively homogeneous areas of the image. Further details on LHE method can be found in Kim et al., (1998).

2.2.3. Contrast-Limited Adaptive Histogram Equalization (CLAHE)

CLAHE is a modification of the LHE method aimed to overcome the problem of over amplification of noise. In this method, the analyst must define a limit for clipping the local histogram and maximum enhancement factor. In the first step, the algorithm divides the input image into non overlapping blocks of equal size. In the second and final step, the algorithm clips the histogram for each block based on the user defined limit, and the cumulative density function is calculated for linearly mapping the pixel intensities. CLAHE method is also computationally intensive. Further details on CLAHE method can be found in Pizer et al. (1990). In this study, $1/8^{th}$ of the image height and weight was set as the window size. In this study, four clipping threshold values (0.01, 0.02, 0.03 and 0.09) were tested.

https://doi.org/10.5194/isprs-archives-XLIV-M-2-2020-45-2020 | © Authors 2020. CC BY 4.0 License.

2.2.4. Linear Contrast Stretch (LCS)

LCS linearly stretches the contrast of an image by spreading the pixel intensity values over the entire dynamic range. The pixel values in low contrast images cover a limited range of intensity values. LCS stretches the intensity values over the entire range of the histogram. After LCS, bright areas in the image become brighter and the dark areas become darker. One of the disadvantages of assigning equal intensity values to both rarely and frequently occurring pixels, some important details might be lost. Also, LCS cannot improve the contrast if the pixel intensity values cover the entire available range. Further details on LCS method can be found in Gillespie (1992).

2.2.5. Gamma Correction (GC)

GC which is also known as Power Law Transform is a spatial domain contrast enhancement method where the value of each pixel in the input image is mapped to a new value based on a transformation function. The transformation function is based on scaling (c) and positive (γ) constants that are specified by the analyst. When γ values are set below 1, narrow range of dark input pixel values are transformed to a wide range of bright output values. When γ values are set above 1, the bright input pixel values are transformed to a narrow range of dark output values. Further details on GC method can be found in Huang et al. (2016). In this study, five potential γ values (0.1, 0.3, 0.5, 0.7, and 0.9) were tested. The value of c was chosen as 1.

2.3 Non-reference Evaluation Metric

Output images generated from each enhancement method was evaluated using the Blind/Reference less Image Spatial Quality Evaluator (BRISQUE). BRISQUE is an assessment metric that calculates the loss in naturalness of the image due to distortions (35). The value of BRISQUE score ranges between 0 and 100 and lower scores indicate better quality output image. Further details on BRISQUE can be found in Mittal et al. (2012).

3. RESULTS

3.1 Window size for LHE

BRISQUE scores of the output images generated with different window sizes are listed in Table 1. Output image generated with a window size of 128 x 128 pixels had the lowest BRISQUE score. Hence this window size was selected to process the rest of the images.

Window size	BRISQUE
8 x 8	43.41
16 x 16	30.84
32 x 32	29.73
64 x 64	30.06
128 x 128	26.98

Table 1. BRISQUE scores of the output images generated with Local Histogram Equalization algorithm using five different window sizes.

3.2 Threshold value for CLAHE

BRISQUE scores of the output images generated with different clipping threshold values are listed in Table 2. The output image generated with a clipping threshold of 0.03 had the lowest BRISUE score of 27.48. Hence this clipping threshold value was selected for processing the rest of the images.

Clipping threshold	BRISQUE
0.01	30.87
0.02	30.86
0.03	27.48
0.09	28.23

Table 2. BRISQUE scores of the output images generated with Contrast-Limited Adaptive Histogram Equalization algorithm with four clipping threshold values.

As the threshold value increased, the contrast of the output image also increased. However, over enhancement problems were noticed when the clipping threshold value of 0.09 was selected.

3.3 Quality Assessment of Output Images

Based on the BRISQUE scores of the output generated from true color Landsat images (Table 3), no one enhancement method consistently outperformed the rest. The mean BRISQUE value for the LHE generated images were lower than those obtained for other enhancement methods (Table 3). BRISQUE scores of the output images generated by the GC method was highest in comparison to other methods.

Image	HE	LHE	CLAHE	LCS	GC
1	19.99	22.17	22.32	30.88	42.19
2	28.77	31.50	33.77	32.46	44.19
3	23.29	28.85	32.30	28.70	43.06
4	32.75	38.82	30.87	39.66	35.41
5	30.37	26.98	22.65	35.29	50.28
6	24.34	28.52	28.72	25.67	35.93
7	26.30	21.28	31.89	31.32	45.84
8	30.47	19.87	33.49	36.34	45.29
9	29.98	31.44	28.20	34.85	46.40
10	32.39	33.29	37.22	31.78	41.61
11	30.68	13.40	30.26	30.88	54.67
12	37.34	12.09	33.98	40.60	48.02
13	38.07	31.18	36.51	40.39	43.73
14	33.86	24.71	31.72	35.16	46.40
15	27.71	7.73	27.72	28.71	30.32
16	30.82	14.85	32.04	25.78	43.95
17	17.28	13.90	27.55	26.81	45.74
18	31.59	27.21	34.35	36.31	42.18
Mean	29.22	23.77	30.86	32.87	43.62
Std. Dev	5.26	8.37	4.00	4.63	5.40

Table 3. BRISQUE scores of the output generated from true color post-flood Landsat images using five enhancement algorithms. Lower BRISQUE score indicates higher output quality.

https://doi.org/10.5194/isprs-archives-XLIV-M-2-2020-45-2020

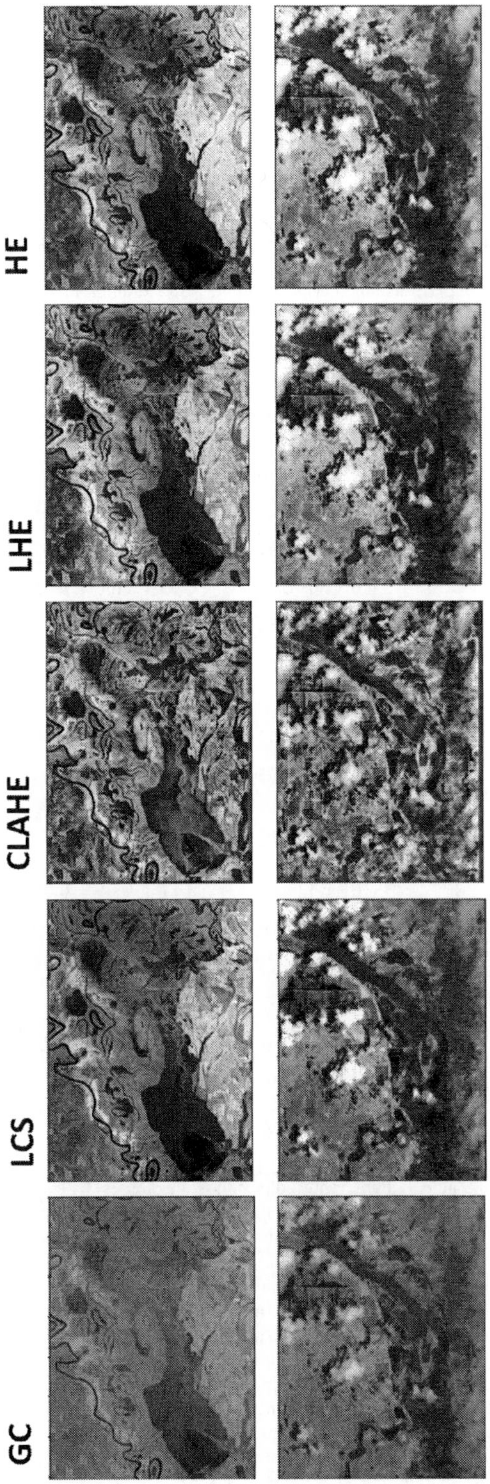

Figure 3. Output generated for two of the sixteen post-flood images using five image enhancement algorithms: Histogram Equalization (HE), Local Histogram Equalization (LHE), Contrast Limited Adaptive Histogram Equalization (CLAHE), Linear Contrast Stretch (LCS), and Gamma Correction (GC).

BRISQUE scores of the output generated from false color images (Table 4) showed a similar pattern. Like the true color images, the mean BRISQUE value for LHE generated images were lower than those obtained for other enhancement methods (Table 4). Among all methods, the output images generated by GC had the highest BRISQUE scores.

Image	HE	LHE	CLAHE	LCS	GC
1	23.02	36.23	24.62	31.96	32.47
2	29.54	17.66	30.57	22.35	36.94
3	30.97	21.61	30.70	29.27	33.62
4	21.86	21.21	24.08	31.26	25.11
5	28.48	23.17	16.13	31.86	39.76
6	24.15	19.13	20.53	18.75	27.28
7	20.79	26.36	30.60	27.50	44.46
8	14.60	22.32	30.98	25.54	36.95
9	30.83	29.76	18.25	28.32	33.92
10	31.55	21.14	33.07	22.17	35.09
11	16.99	13.86	28.06	29.59	40.31
12	23.94	19.26	25.61	32.03	50.83
13	31.00	23.47	29.78	37.25	44.34
14	22.34	23.49	33.39	25.56	46.23
15	27.52	6.88	25.66	21.51	28.52
16	25.19	25.26	25.18	23.65	27.40
17	21.59	25.99	14.65	25.64	32.56
18	33.83	32.59	39.62	32.68	42.87
Mean	25.46	22.74	26.75	27.61	36.59
Std Dev	5.19	6.44	6.28	4.69	7.07

Table 4. BRISQUE scores of the output generated from false color post-flood Landsat images using five enhancement algorithms. Lower BRISQUE scores indicate higher output quality.

Pair-wise comparison (one tailed, paired T-test) of the BRISQUE scores indicate that the output generated from each method was statistically different from the rest (Table 5). Output quality of LHE generated images are significantly different from those generated from other methods.

	LHE	LCS	CLAHE	GC
HE	< 0.01	<0.01	N.S.	<0.001
LHE		<0.001	<0.001	<0.001
LCS			N.S.	<0.001
CLAHE				<0.001

Table 5. Probability values obtained from pair-wise comparison of the BRISUE scores of output images (n = 36) generated by five image enhancement techniques. N.S.: Not significant.

LHE divides the input image into blocks or tiles and redistributes the intensity level of the pixels by computing the histogram of each tile. Hence the overall contrast and edges of the image are enhanced even in the presence of noise, such as clouds (Figure 3). This resulted in lower BRISQUE scores and better overall performance by LHE method. However, this method is sensitive to the window size specified by the analyst.

Output from the HE method had the next lowest BRISQUE scores. Unlike LHE, this method does not require any input parameters such as specifying the window size. However, the pixels in the output images can be saturated due to the clipping effect, which results in loss of information. Hence the BRISQUE scores of HE were higher than that of LHE.

CLAHE method requires the analyst to specify the window size as well as a clip limit. The algorithm limits the contrast of the output image based on those values, which might lead to noise amplification and artifacts along the edges. Areas covered under thin clouds are not distinguishable (Figure). These factors resulted in higher average BRISQUE scores.

LCS increases the contrast of the input images by linearly spreading the histogram over the entire range. As a result, both frequently and rarely occurring pixels are assigned equal intensity values which results in poor quality output image Figure). Hence the output generated by LCS had relatively higher BRISQUE scores than LHE, HE and CLAHE.

Among the five enhancement techniques, GC produced lowest quality images for all images. BRISQUE scores for 36 output images were higher than the corresponding scores obtained for the other 4 methods. The gamma value specified in this study did not spread the histogram over the entire dynamic data range. Based on the results obtained for the 36 images, LHE method generated higher quality images, followed by HE method.

Image	HE	LHE	CLAHE	LCS	GC
1	0.05	0.36	0.06	0.02	0.02
2	0.14	1.33	0.26	0.08	0.07
3	0.05	0.42	0.08	0.02	0.02
4	0.08	0.82	0.22	0.21	0.07
5	0.30	1.07	0.20	0.06	0.07
6	0.12	0.77	0.15	0.05	0.05
7	0.11	0.95	0.16	0.06	0.06
8	0.15	0.78	0.74	0.22	0.11
9	0.18	1.08	0.16	0.09	0.07
10	0.10	0.74	0.19	0.04	0.06
11	0.14	1.00	0.16	0.06	0.11
12	1.82	0.65	0.35	0.02	0.10
13	0.10	0.83	0.19	0.05	0.06
14	0.10	0.72	0.17	0.06	0.08
15	0.10	0.74	0.20	0.06	0.05
16	0.12	0.91	0.13	0.05	0.06
17	0.11	0.82	0.13	0.04	0.05
18	0.13	0.58	0.29	0.03	0.03
Mean	0.22	0.81	0.21	0.07	0.06
Std. Dev	0.39	0.23	0.14	0.05	0.03

Table 6: Time required (seconds/megabytes) by each enhancement algorithm (columns) to process the 18 true color images. LHE required most time in comparison to LCS and GC.

3.4 Processing Time

Standardized time taken (seconds/Megabytes) for processing each true color image was lower for GC and LCS in comparison to the other methods (Table 6). LHE which produced higher quality images, i.e., lower BRISQUE scores, needed most time to process the images.

Standardized time taken (seconds/Megabytes) for processing each false color image was lower for GC and LCS in comparison to the other methods (Table 7). Like the previous batch, LHE which produced higher false quality images, i.e., lower BRISQUE scores, needed more time to process the images.

Image	HE	LHE	CLAHE	LCS	GC
1	0.04	0.30	0.05	0.02	0.02
2	0.13	0.78	0.16	0.05	0.07
3	0.04	0.29	0.05	0.02	0.03
4	0.08	0.51	0.39	0.04	0.03
5	0.11	0.95	0.13	0.06	0.05
6	0.13	0.85	0.13	0.04	0.06
7	0.12	0.88	0.15	0.04	0.05
8	0.15	0.63	0.30	0.04	0.04
9	0.11	1.03	0.18	0.48	0.56
10	0.14	0.75	0.17	0.04	0.06
11	0.10	0.87	0.16	0.12	0.08
12	0.10	0.77	0.20	0.02	0.05
13	0.11	0.68	0.17	0.04	0.05
14	0.13	1.24	0.15	0.04	0.05
15	0.11	0.73	0.15	0.04	0.05
16	0.12	0.82	0.14	0.05	0.05
17	0.12	0.81	0.13	0.04	0.05
18	0.10	0.61	0.29	0.03	0.03
Mean	0.11	0.75	0.17	0.07	0.08
Std. Dev	0.02	0.23	0.07	0.10	0.11

Table 7: Time required (seconds/megabytes) by each enhancement algorithm (columns) to process the 18 false color images. LHE required most time in comparison to LCS and GC.

LHE method divides the input image into tiles and processes each one separately, it takes the most amount of time. Though the CLAHE method also divides the input image into tiles, it did not require the same amount of processing time. Time required by LHE to process images could be a limitation during emergency response, especially when the file size of the input image is large. Under these circumstances HE could be used for enhancing the post-flood satellite images. However, the quality of the output images generated by HE was lower than those generated by LHE. Further studies must process data collected by other remote sensing satellites for other regions with different characteristics.

4. CONCLUSIONS

Based on the BRISQUE scores of the output images, LHE algorithm will be suitable for enhancing post-flood images covered by thin-clouds and haze. There was no difference in quality of the output generated from the true color or false color images. However, LHE method needed most time to process the images.

If numerous images have to be rapidly processed, HE method could serve as the alternative method. The quality of the output generated from HE algorithm was slightly lower than those generated by the LHE algorithm.

LCS and GC algorithms needed relatively the least amount of time to process the images however their output quality was much lower. The output images generated for both true and false color images had higher BRISQUE scores.

CLAHE required relatively less time than LHE, the quality of the output images was lower due to the presence of artifacts. The methods that required relatively less processing time did not generate quality output.

ACKNOWLEDGEMENTS

Authors thank the US Geological Survey for proving Landsat data at no-cost and Dr. K.P. Soman, Head, Center for Computational Engineering and Networking at Amrita Vishwa Vidyapeetham (Deemed University) for his invaluable support.

REFERENCES

Bidwai, P., Tuptewar, D.J., 2015: Resolution and contrast enhancement techniques for grey level, color image and satellite image, *International Conference on Information Processing (ICIP), Pune*, 511-515, doi.org/10.1109/INFOP.2015.7489437.

Brivio, P.A., Colombo, R., Maggi,M., Tomasoni R., 2002: Integration of remote sensing data and GIS for accurate mapping of flooded areas. *International Journal of Remote Sensing* 23(3), 429-441. doi.org/10.1080/01431160010014729.

Chen, S., Ramli, A.R., 2003: Minimum mean brightness error bi-histogram equalization in contrast enhancement, *IEEE Transactions on Consumer Electronics*, 49(4), 1310-1319, doi.org/10.1109/TCE.2003.1261234.

Gianinetto, M., Villa, P., Lechi G., 2005: Postflood damage evaluation using Landsat TM and ETM+ data integrated with DEM. *IEEE Transactions on Geoscience and Remote Sensing,* 44(1), 236-243. doi.org/10.1109/TGRS.2005.859952.

Gillespie, A.R., 1992: Enhancement of multispectral thermal infrared images: Decorrelation contrast stretching. *Remote Sensing of Environment*, 42(4), 147-155. doi.org/ 10.1016/0034-4257(92)90098-5

Kerele, N., Oppenheimer, C., 2002: Satellite remote sensing as a tool in lahar disaster management. *Disasters*, 26(2): 140-160. doi.org/10.1111/1467-7717.00197.

Kussul, N., Shelestov, A., Skakun S., 2008: Grid system for flood extent extraction from satellite images. Earth Science Informatics 1, 105-117 (2008). doi.org/10.1007/s12145-008-0014-3.

Huang, Z., Zhang, T., Li, Q., Fang, H., 2016: Adaptive gamma correction based on cumulative histogram for enhancing near-infrared images. *Infrared Physics & Technology*, 79, 2015-215. doi.org/ 10.1016/j.infrared.2016.11.001.

Kim, T.K., Paik, J.K., Kang, B.S., 1998: Contrast enhancement system using spatially adaptive histogram equalization with temporal filtering, *IEEE Transactions on Consumer Electronics*, 44 (1), 82-87, doi.org/10.1109/30.663733.

Lee, K., Lin C., 2016: Cloud detection of optical satellite images using support vector machine. *Int. Arch. Photogramm. Remote Sens.,* XLI-B7, 289-293. doi.org/10.5194/isprsarchives-XLI-B7-289-2016.

Luscombe, B.W., Hassan, H.M., 1993: Applying remote sensing technologies to natural disaster risk management: Implications for developmental investments. *Acta Astronautica*, 29(10-11), 871-876. doi.org/10.1016/0094-5765(93)90169-W.

Mittal, A., Moorthy, A.K., Bovik, A.C., 2012: No-Reference Image Quality Assessment in the Spatial Domain. *IEEE Transactions on Image Processing*, 21 (12), 4695-4708, do.org/10.1109/TIP.2012.2214050.

Moniruzzaman, M., Shafuzzaman, M., Hossain, M.F., 2014: Brightness preserving Bi-histogram equalization using edge pixels information, *2013 International Conference on Electrical Information and Communication Technology (EICT)*, Khulna, 1-5, doi.org/10.1109/EICT.2014.6777872.

Pizer, S.M., Amburn, E.P., Austin, J.D., Cromartie, R., Geselowitz, A., Greer, T., Romeny, B.T., Zimmerman, J.B., Zuidervel, K., 1987: Adaptive histogram equalization and its variations. *Computer Vision, Graphics, and Image Processing*, 39(3): 355-368. doi.org/10.1016/S0734-189X(87)80186-X.

Pizer, S.M., Johnston, R.E., Ericksen, J.P., Yankaskas, B.C., Muller, K.E., 1990: Contrast-Limited Adaptive Histogram Equalization: Speed and Effectiveness. Proceedings of the First Conference on Visualization in Biomedical Computing, Atlanta, GA, USA, 337-345. doi.org/ 10.1109/VBC.1990.109340.

Singh, H., Kumar, A., Balyan, L.K., Singh, G.K., 2019: A novel optimally weighted framework of piecewise gamma corrected fractional order masking for satellite image enhancement, *Computers & Electrical Engineering*, 75, 245-2661. doi.org/ 10.1016/j.compeleceng.2017.11.014.

Startk J.A., 2000: Adaptive image contrast enhancement using generalizations of histogram equalization, *IEEE Transactions on Image Processing*, 9(5), 889-896, doi.org/10.1109/83.841534.

Wang, Y., Chen, Q., Zhang, B., 1999: Image enhancement based on equal area dualistic sub-image histogram equalization method, *IEEE Transactions on Consumer Electronics*, 45(1), 68-75, doi.org/10.1109/30.754419.

DEEP LEARNING FOR REMOTE SENSING IMAGE CLASSIFICATION FOR AGRICULTURE APPLICATIONS

Leila Hashemi-Beni[1*], Asmamaw Gebrehiwot[2]

[1*] Geomatics Program, Department of Built Environment, North Carolina A&T State University - lhashemibeni@ncat.edu
[2] Applied Science and Technology Ph.D. Program, Department of Built Environment, North Carolina A&T State University - aagebrehiwot@aggies.ncat.edu

KEY WORDS: Precision Agriculture, Remote Sensing, U-Net, FCN-8s, Deep Learning, Image Segmentation

ABSTRACT:

This research examines the ability of deep learning methods for remote sensing image classification for agriculture applications. U-net and convolutional neural networks are fine-tuned, utilized and tested for crop/weed classification. The dataset for this study includes 60 top-down images of an organic carrots field, which was collected by an autonomous vehicle and labeled by experts. FCN-8s model achieved 75.1% accuracy on detecting weeds compared to 66.72% of U-net using 60 training images. However, the U-net model performed better on detecting crops which is 60.48% compared to 47.86% of FCN-8s.

1. INTRODUCTION

Weed controlling is one of the main problems that farmers must deal with. Weeds profoundly affect farm productivity by invading crops and smother pastures and significantly decrease the quality of the harvested crops (Milberg et al., 2004). Herbicides are widely used globally to enhance food production; however, it can cause harm to the environment and the ecosystem (Horrigan et al., 2002). In the traditional weeds control approaches, the herbicide is applied over the whole field, even for the area without weeds where no treatment is required. Precision agriculture techniques should regularly monitor crop growth to maximize yield while minimizing the use of resources such as chemicals and reducing the side effects of herbicides on the environment (Duckett et al., 2018). Thus, accurate weeds detection and mapping, and their local treatment are essential steps to improve weed and crop control in modern agriculture.

Recently, unmanned aerial systems (UAVs) have become suitable platforms for acquiring data for crop and weeds monitoring. UAVs can acquire high-resolution imagery at a low cost, and less dependence on weather conditions (Hashemi-Beni et al., 2018, Gebrehiwot et al., 2019, Vinh et al., 2019).

Several methods have been proposed for weed recognition over the last decades. Vegetation index, such as the normalized difference vegetation index (NDVI) is one of the methods to segment weeds in an agricultural field (Dyrmann et al., 2014). The main challenge of this approach is dealing with overlapping plants to separate weeds and crops. Texture-based models have shown great performance in detecting and discriminating plants from images with overlapping leave (Pahikkala et al., 2015). Machine learning-based approaches have gained attention for detecting weeds and crops (Murawwat et al., 2018). Murawwat, et al. (2018) use a support vector machine (SVM) classifier to classify carrot crops and weeds. They used 72 samples for training and eight samples for testing and achieved more than 50% of classification accuracy. The challenge of the traditional ML approach such as SVM or RF classifiers is that feature extraction is not automatic and handcrafted features generation is a time-consuming stage. Some studies prove that unlike conventional machine learning methods, deep learning can efficiently deal with the limitations of handcrafted features for classifying weed and crops by extracting the features directly from the input data (Lee et al., 2015). Recently, there has been considerable progress in the classification of remote sensing data using deep learning for different applications including agricultural tasks. Some studies used a convolutional neural network (CNN) in agricultural applications such as weed and crop classification (Mortensen et al., 2016, Potena et al., 2016, Di Cicco et al., 2017). Mortensen et al. (2016) used the VGG-16 CNN model to classify weeds using mixed crops of an oil radish plot with barely, weed, stump, grass, and background soil images. Potena et al. (2016) presented a perception system for weed crop classification that uses shallow and deeper CNNs. The shallow CNN is used to detect vegetation, while the deeper one is used to classify weeds and crops. Di et al. (2017) used a SegNet to procedurally generate large synthetic training datasets randomizing the key features of the target environment (i.e., crop and weed species, type of soil, light conditions). Hashemi-Beni, et al. (2020) applied the u-net model to detect and discriminate crop and weeds using a small dataset. They applied data augmentation techniques such as random cropping, random rotation, and reflection to improve the classification results.

This research provides a comparison of the U-Net and FCN-8s models for segmenting weeds from Crop/Weed Field Image Dataset (CWFID), which was introduced in Haug et al. (2014).

2. DATASET AND ANNOTATION

The CWFID (Haug et al., 2014) was used to train and test the U-Net classifier in this study. The CWFID dataset comprises field images in top-down view that were collected with an autonomous field robot Bonirob in an organic carrot farm in 2013. The CWFID includes 60 images with a size of 1296 x 966 pixels. All data acquisition was carried out during field tests, and images were acquired while the crop was in the growth stage and one or more true leaves were present. These images were labeled with a vegetation segmentation mask and crop/weed labeled image (In total: 162 and 332 crop and weed plants, respectively). A sample image of this dataset is shown in Figure 1. The CWFID was downloaded from https://github.com/cwfid.

This contribution has been peer-reviewed.
https://doi.org/10.5194/isprs-archives-XLIV-M-2-2020-51-2020 | © Authors 2020. CC BY 4.0 License.

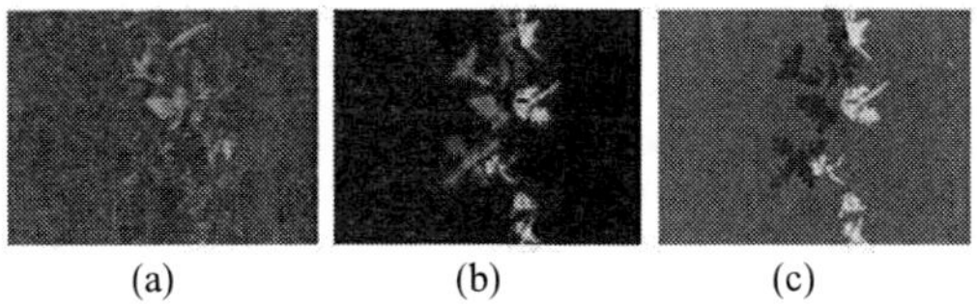

Figure 1. Sample training image. (**a**) training image; (**b**) labeled image, (**c**) the image relabeled using MATLAB tool.

Figure 1 shows one sample example of the training and its corresponding labeled image. The black, green, and red colors represent soil, crops, and weeds, respectively.

3. METHODS AND DATA PROCESSING

3.1 Network Architecture

3.1.1 U-net Architecture: The U-Net is a type of CNN architecture proposed by Ronneberger et al. (2015) for biomedical image segmentation. The U-Net architecture is based on the fully convolutional network, and it was modified to give accurate classification results with fewer training data. As shown in Figure 2, the architecture of U-Net looks like a letter ‘U’ that justifies its name.

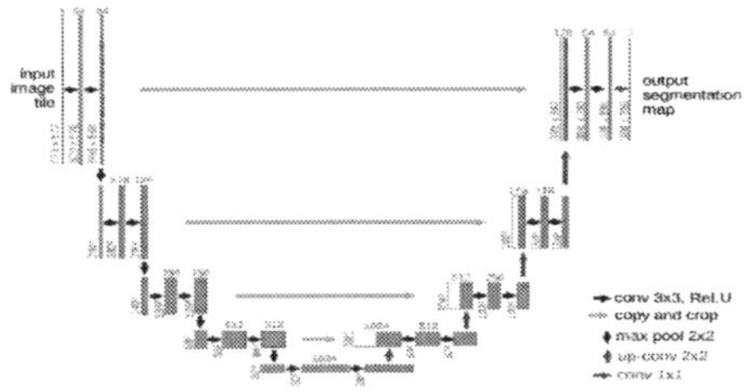

Figure 2. U-Net architecture (Ronneberger et al., 2015)

The U-Net architecture has two structures: a shrinking path (left side) and an expanding path (right side). The shrinking path (also called the encoder) used to extract and capture the context in the image and the expanding path combines the feature maps. Then, more precise predictions of the pixel points on the edge can be obtained. The encoder is just a traditional stack of convolutional and max-pooling layers. The expanding structure (or decoder) is used to enable precise localization using transposed convolutions. Thus, it is an end-to-end fully convolutional network that only contains convolutional layers and does not contain any dense layer because of which it can accept the image of any size.

3.1.2 FCN-8s architecture: The FCN was proposed by Long et al. (2015) to train an end-to-end for semantic segmentation. In this model, VGG16 fully connected based classification layers were replaced by convolutional layers to maintain the 2-D structure of images. VGG-16 is a CNN architecture proposed by Simonyan et al. (2014) to investigate the effect of the convolutional network depth on its accuracy in the large-scale image recognition setting. As shown in Figure 3, because of the gradual upsampling of the scoring layer and merging of features from earlier layers, a fine label map is obtained using the FCN-8s. For this study, a fully convolutional neural network (FCN) with a stride of 8 (FCN-8s) model was fine-tuned. The FCN-8s is composed of locally connected layers, such as convolution, pooling, and upsampling, without having any dense layer. This allows reducing the number of parameters and computation time. Given that all connections are local, FCN-8s can work on any image size.

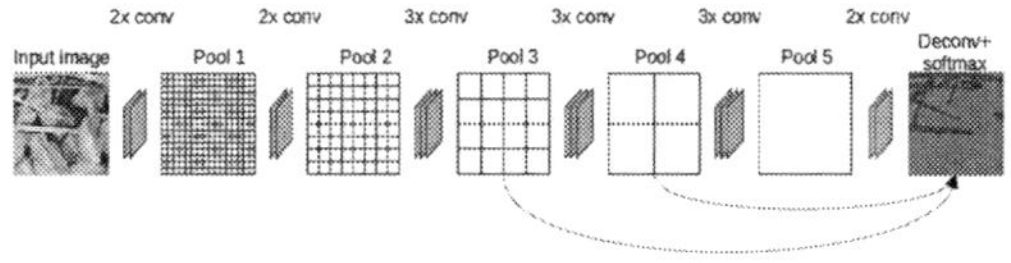

Figure 3. FCN-8s architecture (Skovsen et al., 2017)

3.2 Training U-Net and FCN-8s

In this study, the U-net and FCN-8s models fine-tuned and trained using Stochastic Gradient Descent (SGD) with a learning rate of 0.001, and a maximum epoch of 16. A 10-fold cross-validation method was used to estimate the ability of the U-net and FCN models on unseen data. For this purpose, we randomly partitioned the training images into 10 equal parts. At each run, the union of 9 parts was put together to form a training set, and the remaining 1-part used as a test set to estimate the classification errors. The above steps are repeated ten times, using a different fold as the testing set each time. Finally, the mean error from all folds was used to estimate the potential of the U-net and FCN-8s models. It took approximately 8 hours to cross-validating U-Net and FCN-8s using a single GPU (NVIDIA Quadro M4000).

4. RESULTS AND DISCUSSION

In this study, a confusion matrix was used to analyze the classification results of the U-net and FCN-8s models. The qualitative classification results of U-Net and FCN-8s are shown in Figure 4, and the detailed results on how the U-Net and FCN-8s models performed for each class (for soil, weeds, and crops classes) are described in Table 1 and Table 2, via the confusion matrices. The U-Net and FCN-8s models achieved an overall accuracy of 75.2% and 72.1%, respectively for separating weeds from crops.

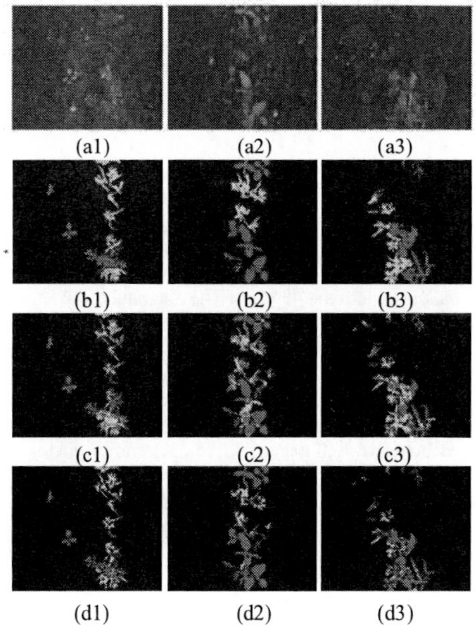

Figure 4. Classification results of U-Net and FCN-8s. (a1-a3) sampe test images; (b1-b3) Ground truth, and (c1-c3) U-net classification results; (d1-d3) FCN-8s results

https://doi.org/10.5194/isprs-archives-XLIV-M-2-2020-51-2020

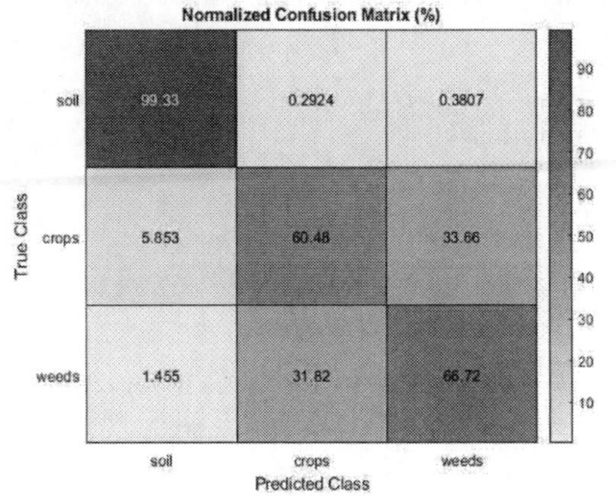

Table 1. Confusion matrix of U-Net Classification (unit-percentage)

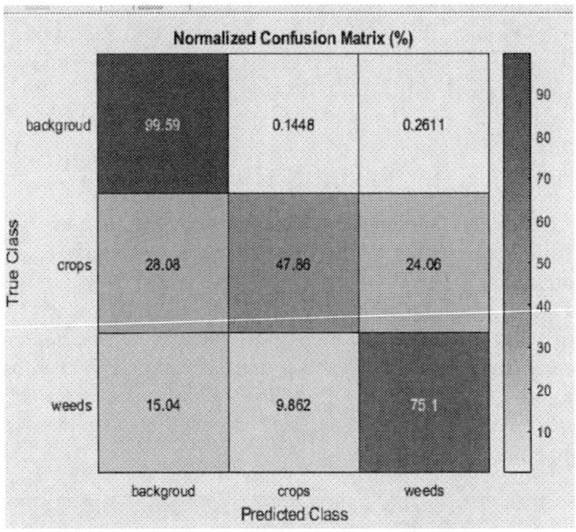

Table 2. Confusion matrix of FCN-8s Classification(unit-percentage)

Table 1 and Table 2 show the correspondence between the predicted and the validation values. The cells of a confusion matrix represent the percentage of correct and incorrect predictions for all the possible correlations. The cell in the ith row and *j*th column means the percentage of the *i*th class samples, which classified into the *j*th class. The diagonal cell of the matrix contains the number of correctly identified pixels for each class. As the tables demonstrate, the U-Net model achieved the classification accuracy of 99.3% for soil, 60.48% for crops, and 66.72% for weeds, whereas FCN-8s an accuracy of 99.59% for soil, 47.86% for crops, and 75.1% for weeds. The U-net model achieved better overall accuracy (75.2%) compared to FCN-8s (72.1%). However, FCN-8s showed better performance on extracting weeds (75.1%) compared to U-net (66.72%). The results also show that about 33% and 24% of weeds in U-net and FCN-8s, respectively were misclassified as crops. This is because weeds and crops have a similar spectral response, which makes it hard to separate them using U-Net and FCN-8s classifiers solely from optical imagery.

5. CONCLUSION

Mapping the location of the weeds and locally treat those areas are essential to improve weed control capacity. This study compared the performance of two well-known deep learning-based methods to distinguish crops from weeds using high-resolution imagery. The U-Net and FCN-8s models were fine-tuned to classify the CWFID into three classes (soil, crops, and weeds). The fine-tuning and transfer learning technique allowed us to overcome the problem of small dataset size. The U-Net and FCN-8s models achieved an overall accuracy of 75.2% and 72.1%, respectively for classifying the CWFID dataset into three classes using 60 training images. In future research, we will incorporate crop geometry constrains to the model to improve classification accuracy.

ACKNOWLEDGEMENT

This material is based upon work supported by the National Science Foundation under Grants No. 1832110 and No.1800768.

REFERENCES

Di Cicco, M., Potena, C., Grisetti, G., Pretto, A., 2017. Automatic model based dataset generation for fast and accurate crop and weeds detection. In *2017 IEEE/RSJ International Conference on Intelligent Robots and Systems (IROS)*, IEEE, 5188-5195.

Duckett, T., Pearson, S., Blackmore, S., Grieve, B., Chen, W.H., Cielniak, G., Cleaversmith, J., Dai, J., Davis, S., Fox, C., From, P., 2018. Agricultural robotics: the future of robotic agriculture. *arXiv preprint arXiv:1806.06762*.

Dyrmann, M., Christiansen, P., 2014. Automated Classification of Seedlings Using Computer Vision: Pattern Recognition of Seedlings Combining Features of Plants and Leaves for Improved Discrimination. Aarhus University.

Gebrehiwot, A., Hashemi-Beni, L., Thompson, G., Kordjamshidi, P., Langan, T. E., 2019. Deep convolutional neural network for flood extent mapping using unmanned aerial vehicle data. *Sensors*, *19*(7), 1486.

Hashemi-Beni, L., Gebrehiwot, A., 2020. Weed and crop discrimination using U-net learning IGRASS.

Hashemi-Beni, L., Jones, J., Thompson, G., Johnson, C., Gebrehiwot, A., 2018. Challenges and opportunities for UAV-based digital elevation model generation for flood-risk management: A case of Princeville, North Carolina. *Sensors*, *18*(11), 3843.

Hashemi-Beni, L, et al. Weed and crop discrimination using U-net learning, IGRASS 2020, In Press.

Haug, S., Ostermann, J., 2014. A crop/weed field image dataset for the evaluation of computer vision based precision agriculture tasks. In *European Conference on Computer Vision*, 105-116. Springer, Cham.

Horrigan, L., Lawrence, R. S., Walker, P., 2002. How sustainable agriculture can address the environmental and human health harms of industrial agriculture. *Environ Health Perspect, 110*(5), 445-456.

Lee, S. H., Chan, C. S., Wilkin, P., Remagnino, P., 2015. Deep-plant: Plant identification with convolutional neural networks. In *2015 IEEE international conference on image processing (ICIP)*, IEEE, 452-456.

Long, J., Shelhamer, E., Darrell, T., 2015. Fully convolutional networks for semantic segmentation. In *Proceedings of the IEEE conference on computer vision and pattern recognition*, 3431-3440.

Milberg, P., Hallgren, E., 2004. Yield loss due to weeds in cereals and its large-scale variability in Sweden. *Field Crops Research*, *86*(2-3), 199-209.

Mortensen, A. K., Dyrmann, M., Karstoft, H., Jørgensen, R. N., Gislum, R., 2016, Ju. Semantic segmentation of mixed crops using deep convolutional neural network. In *Proc. of the International Conf. of Agricultural Engineering (CIGR).*

Murawwat, S., Qureshi, A., Ahmad, S., Shahid, Y., 2018. Weed Detection Using SVMs. *Engineering, Technology & Applied Science Research*, *8*(1), 2412-2416.

Pahikkala, T., Kari, K., Mattila, H., Lepistö, A., Teuhola, J., Nevalainen, O. S., Tyystjärvi, E., 2015. Classification of plant species from images of overlapping leaves. *Computers and Electronics in Agriculture*, *118*, 186-192.

Potena, C., Nardi, D., Pretto, A., 2016. Fast and accurate crop and weed identification with summarized train sets for precision agriculture. In *International Conference on Intelligent Autonomous Systems*, 105-121. Springer, Cham.

Ronneberger, O., Fischer, P., Brox, T., 2015. U-Net: Convolutional Networks for Biomedical Image Segmentation. *Lecture Notes in Computer Science*, 9351, 234-241.

Simonyan, K., Zisserman, A., 2014. Very deep convolutional networks for large-scale image recognition. *arXiv preprint arXiv:1409.1556*.

Skovsen, S., Dyrmann, M., Mortensen, A. K., Steen, K. A., Green, O., Eriksen, J., Karstoft, H., 2017. Estimation of the botanical composition of clover-grass leys from RGB images using data simulation and fully convolutional neural networks. *Sensors*, *17*(12), 2930.

Vinh, K., Gebreyohannes, S., Karimoddini, A., 2019. An Area-Decomposition Based Approach for Cooperative Tasking and Coordination of UAVs in a Search and Coverage Mission. In *2019 IEEE Aerospace Conference*, IEEE, 1-8.

THE ACCESSIBILITY AND SPATIAL PATTERNS OF GREEN OPEN SPACE BASED ON GIS

Danxuan Liu [1, *], Yishao Shi [1]

[1] College of Surveying and Geo-Informatics, Tongji University, Shanghai, China - (liudanxuan, shiyishao)@tongji.edu.cn

KEY WORDS: Green Open Space (GOS), Availability, Spatial Patterns, Coverage Rate, Morphological Features, GIS

ABSTRACT:

Studies show that the green open space (GOS) is beneficial to visitors' mental and physical health and has positive social values. This study took four global cities as examples, namely Shanghai, Tokyo, New York and London. The per capita area, the coverage rate and the availability of GOS were calculated in this study. Then the GOS was classified according to the scales and morphological features. And the author analyzed the relations between availability and spatial patterns. The results showed that the four cities could be classified into two classes. Shanghai and Tokyo are high-population-density cities with medium GOS coverage and availability, and New York and London are medium-population-density cities with high GOS coverage and availability. It was found that the high GOS coverage rate did not necessarily lead to a higher availability. Shanghai and London could increase the amount of small GOS to ease the shortage of availability. And London and Tokyo could consider adding linear GOS to improve the connectivity of GOS.

1. INTRODUCTION

The GOS is one of the important elements of a city. It provides venues to citizens for entertainment and leisure and has certain social function. Frequent visits to GOS are beneficial to mental and physical health. Moreover, it possesses great ecological service values.

People are more likely to benefit from the GOS with high quality. The remote-assessment method is able to evaluate the quality of a large number of open space without direct observation (Taylor et al., 2011). It is found that the quality of open space in deprived urban areas cannot satisfy the residents, and the inequality exists in some areas (Hoffimann et al., 2017; Abbasi et al., 2016). For one thing, the higher quality and percentage of open space probably lead to better quality of life (Ambrey and Fleming, 2013; Douglas et al., 2018); for another, the area of open space and the distance to open space will influence the price of a house (Wu et al., 2015; Cho et al., 2010). There is a study demonstrating that the abundant vegetation in open space relates to the lower crime rate (Wolfe and Mennis, 2012). Furthermore, urban green infrastructure is beneficial to the urban sustainability and ecological health (Lovell and Taylor, 2013).

Kabisch et al. (2016) calculated the availability of urban green space based on the land cover data and population data grid. Xu et al. (2018) assessed the impact of urban dynamics on the availability of green space under different scenarios at both regional and sub-regional level. Cetin (2015) found that the distance and the accessibility of green space were closely related to the provision of recreational needs in Kutahya. Wu et al. (2017) used gravity-based accessibility by park type and then explored the relationships between accessibility to parks and housing prices using a hedonic price model. Kim and Nicholls (2017) explored the accessibility and equity of open space by using the geographically weighted regression in Detroit Metropolitan Area. Kang (2015) investigated the impact of accessibility to land use on walking in Seoul by using four newly developed accessibility indices.

Thus far, there is no research to study the relationship between availabilities and spatial patterns of GOS. This study calculated the availabilities and analysed the spatial patterns of four cities, and then investigated the correlation between them. The remainder of the paper is organized as follows. The first section introduces the background information of the study areas. The second section presents the data sources and the research methods. The third section explains the results of GOS indices, the availability and the spatial patterns. The fourth section discusses the relations between the availability and the distribution of GOS and assesses the impacts of spatial pattern of GOS. In the final section, the study summarizes the results and suggests policy implications.

2. STUDY AREAS: SHANGHAI, TOKYO, NEW YORK AND LONDON

This study selected the four global cities Shanghai, Tokyo, New York and London as examples for the following reasons. According to the world city ranking 2018 (published by GaWC), London and New York are the Top 2 world cities and classified as Alpha ++. Shanghai and Tokyo which are in Alpha+ level, are two of the most competitive metropolises in Asia. While Shanghai and Tokyo are high-density cities, which is different from New York and London. We can make the comparison of the GOS distribution in different global cities by conducting analysis in the foregoing four cities.

The study areas of the four cities are the central city of Shanghai, the special wards of Tokyo, New York City and London postal district (They are called Shanghai, Tokyo, New York and London below). These four study areas are central areas of their respective metropolitan areas. And they have similar land areas, which makes the four global cities more comparable. The unit of analysis of the central city of Shanghai is subdistrict. In this study, the subdistricts whose governments are within the Shanghai outer ring expressway constituted the study region. The London postal district is comprised of the E, EC, N, NW, SE, SW, W and WC postcode areas. The unit of analysis of London postal district is

* Corresponding author

This contribution has been peer-reviewed.
https://doi.org/10.5194/isprs-archives-XLIV-M-2-2020-55-2020 | © Authors 2020. CC BY 4.0 License.

wards. If more than half of one ward's land area is within the London postal district, it was included in the study area. The unit of analysis of the special wards of Tokyo and New York City are district (chome) and neighborhood. The study areas of Shanghai, Tokyo, New York and London comprise 104, 3184, 359 and 194 units respectively. The populations and areas of the four cities are shown in table 1.

City	Population	Area (km^2)	Unit of Analysis
Shanghai	11061971	683.88	104
Tokyo	9272740	628.54	3184
New York	8175133	783.12	359
London	4874271	624.54	194

Table 1. Basic information of the study areas

3. DATA AND METHODS

The three indices which are per capita area, coverage rate and availability of GOS were calculated in this research. The availability is the proportion of citizens who can access the GOS, and it was calculated on the basis of the methods proposed by Kabisch et al. (2016).

3.1 Data Sources and Pre-processing

The research data in this study includes the spatial data of population (Figure 1) and GOS of the four study areas. The demographic data of the four cities was acquired from their official census data. And the data sources of their GOS data are different. The vector data of GOS of Shanghai was outlined by the author based on the digital map of tianditu. To be specific, the author added the Tianditu from WMS (Web Map Service) and a new polygon shapefile in QGIS, and then outlined the GOS in the shapefile by adding new features. The data of Tokyo and London is the vector data of open street map which is exported from Hotosm. The data of New York City is downloaded from the website of NYC OpenData. The detailed information of the data sources is described in the table 2.

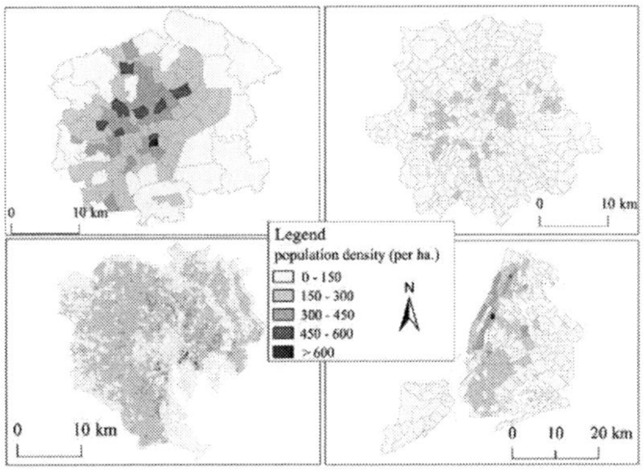

Figure 1. The population density of the four study areas

3.2 The per capita area and coverage rate of GOS

The two GOS indices including per capita area and coverage rate were calculated in this study. The per capita area is the quotient of dividing the total area by the population. And the coverage rate is the ratio of the area of GOS to the area of the study region.

City	Data Sources
Shanghai	Demographic data: Sixth National Population Census of China (year 2010) GOS data: Tianditu (year 2019, http://www.tianditu.gov.cn/)
Tokyo	Demographic data: National Population Census of Japan 2015 GOS data: open street map exported from Hotosm (year 2019, https://export.hotosm.org/en/v3/)
New York	Demographic data: New York City Population By Neighborhood Tabulation Areas (year 2010,https://data.cityofnewyork.us/City-Government/New-York-City-Population-By-Neighborhood-Tabulatio/swpk-hqdp) GOS data: the data provided by NYC Open Data (year 2018, https://data.cityofnewyork.us/Recreation/Open-Space-Parks-/g84h-jbjm)
London	Demographic data: 2011 Census Demography (https://data.london.gov.uk/dataset/981e9136-a06a-44ec-a067-10f3d786cd3f) GOS data: open street map exported from Hotosm (year 2019, https://export.hotosm.org/en/v3/)

Table 2. Data sources of demographic and GOS data

3.3 The availability of GOS

Firstly, this paper determined the accessibility standard. The four cities in this study have different accessibility standards (table 3). This study synthesized the four cities' accessibility standards and developed a new standard to compare the accessibility of the four cities. In Europe, the European Environment Agency (EEA) recommends that people should have access to green space within 15 min walking distance (Barbosa et al., 2007). People's walking velocity is 1.3m/s to 1.4m/s (Ji and Pachi, 2005), so the upper limit set for the service distance is 1200 meters. In addition, the smallest GOS in the four cities' accessibility standard is 0.25 hectares, so 0.25 hectares is the minimum land area which has the shortest service distance 250 meters. And the availability of GOS smaller than 0.25 hectares is not calculated.

Secondly, the author created the buffer zones for the GOS in QGIS. In order to determine the region having access to the GOS, the author created the buffer zones for the GOS according to the accessibility standard. This work was done in QGIS by using the "buffer" function in processing toolbox.

Thirdly, this paper calculated the availability. The availability is calculated based on the buffer zones and the population data. In QGIS, the population data intersected with buffer zones of GOS using the "intersection" in the vector geoprocessing tools and then the buffer zones in each unit of analysis are obtained. Supposing one study urban area C have n units of analysis, the number of citizens in one unit of analysis that have access to the GOS is given by:

https://doi.org/10.5194/isprs-archives-XLIV-M-2-2020-55-2020

$$Q_i = P_i \frac{B_i}{A_i} \quad i=1,2,......,n \tag{1}$$

where P_i = citizen number of the analysis unit i
B_i = intersection land area of the analysis unit i with the buffer zones
A_i = land area of the analysis unit i.

The availability of the study urban area C is given by:

$$R(C) = \frac{\sum_{1}^{n} Q_i}{\sum_{1}^{n} P_i} \quad i = 1,2,......,n \tag{2}$$

City	Reference Standard	Size of Open Space and Service Distance
Shanghai	Technical criteria for regulatory planning of Shanghai (amended version 2016)	0.04 hectares, 300-500 meters; 0.3 hectares, 500-1000 meters; 4 hectares, 2000 meters; 10 hectares, 5000 meters
Tokyo	City planning park · the improvement policy of green space (amended version)	2500 square meters, 250 meters; 10 hectares, 2 kilometers
New York	National recreation & park association (NPRA) parks and open space guidelines	1 arce, 1/4 mile; 5-10 arces, 1/4-1/2 mile; 30-50 arces, 1/2-3 miles; > 50 arces, services the entire community.
London	The London Plan March 2016	2 hectares, 400 meters; 20 hectares, 1.2 kilometers; 60 hectares, 3.2 kilometers; 400 hectares, 3.2-8kilometers

Table 3. The availability standards of the four cities

4. RESULTS

4.1 The per capita area and coverage rate of GOS

The total area and per capita area of GOS are shown in table 4. For New York, it has the highest area and coverage rate and the second-highest per capita area of GOS. The coverage rate and per capita area of New York is 12.18% and 11.65 m^2. London is next to New York in area and coverage rate, and ranks first in per capita area. London's per capita area is 15.21 m^2 and its coverage rate is 10.83%. The values of Shanghai and Tokyo are inferior to those of New York and London. The per capita area and coverage rate of Shanghai and Tokyo are less than half of London's and New York's.

4.2 The availability of GOS

Figure 2 and Figure 3 show the area and number of GOS based on the classification standard (table 5) in 3.3.1. There are numerous GOS less than 0.25 ha or larger than 20 ha in New York. Tokyo has the maximum total area of the GOS less than 0.25 ha among the four cities, while the area and number of GOS larger than 20 ha in Tokyo are less than other three cities. London performs better than Shanghai and Tokyo in the GOS larger than 20 ha, while the area and number of GOS less than 0.25 ha in London is less than New York's and Tokyo's. Shanghai needs to improve on the GOS less than 0.25 ha and larger than 20 ha. There is no significant difference in the other three classes of GOS for the four cities.

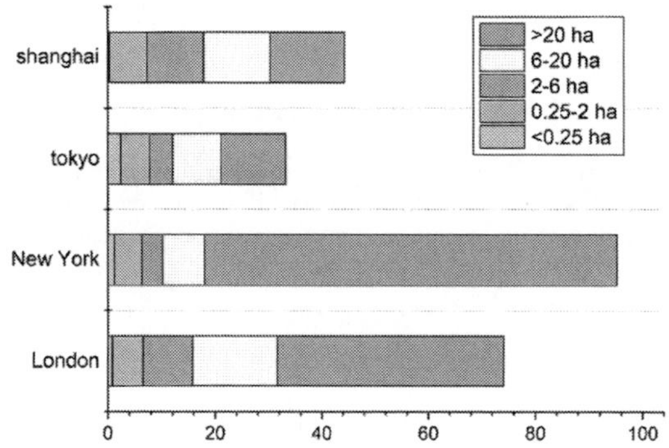

Figure 2. The area of green open space

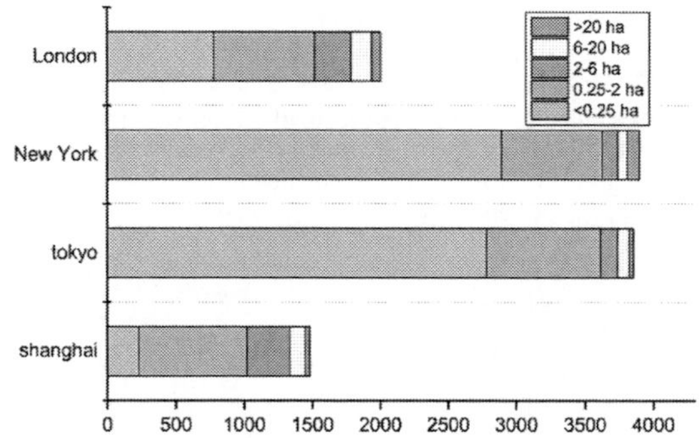

Figure 3. The number of green open space

City	Area of GOS (km^2)	Per Capita Area of GOS(m^2)	Coverage Rate
Shanghai	44.45	4.02	6.50 %
Tokyo	33.19	3.58	5.28 %
New York	95.21	11.65	12.16 %
London	74.12	15.21	11.87 %

Table 4. The per capita area and coverage rate of GOS

The Figure 4 shows the buffer zones which were created based on the methods in 3.3.2. The buffer zones of London form continuous area. In New York, the buffer zones in Manhattan, The Bronx and Staten Island are continuous, while there are both

https://doi.org/10.5194/isprs-archives-XLIV-M-2-2020-55-2020 | © Authors 2020. CC BY 4.0 License.

continuous areas and scattered small areas in Queens and Brooklyn. For Shanghai and Tokyo, the spatial feature of the buffer zone is circle distribution. The large buffer zones form the continuous circular areas and the small ones scatter between the two neighboring circles.

Size of Open Space	Service Distance
> 20 hectares	1200 meters
6-20 hectares	800 meters
2-6 hectares	400 meters
0.25-2 hectares	250 meters

Table 5. The accessibility standard

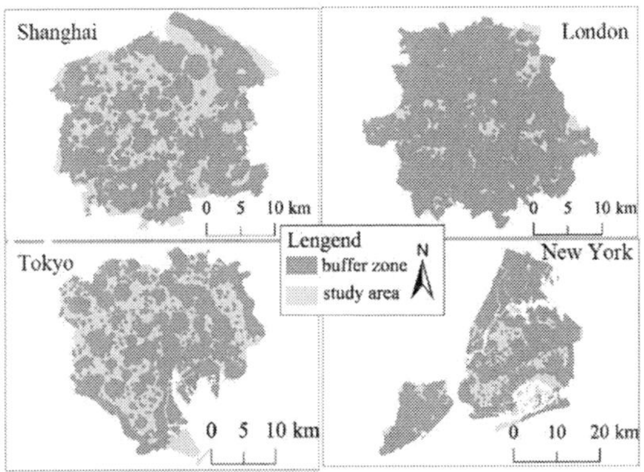

Figure 4. The buffer zones of the four study areas

The table 6 shows the availability of GOS. Among the four cities, London has the highest availability of 91.73%, and the availability of New York is 80.45% which is next to London's. While Shanghai and Tokyo have wide gaps with London and New York. The availability of Shanghai and Tokyo are 64.64% and 61.63%, which means that less than two-thirds of their citizens have access to the GOS larger than 0.25 ha.

City	Availability (%)
Shanghai	64.64
Tokyo	61.63
New York	80.45
London	91.73

Table 6. The availabilities of the study areas

4.3 The spatial pattern of GOS

According to the population density, GOS coverage and availability, we classified the four cities into two classes: Shanghai and Tokyo are the high-population-density cities with medium GOS coverage and availability, and New York and London are the medium-population-density cities with high GOS coverage and availability. Besides, based on the scales we classified the GOS into three classes: small (< 0.25 ha), medium (0.25 ha – 20 ha) and large (> 20 ha). In the aspect of small GOS, Shanghai is the worst among the four cities and London is slightly better than Shanghai. In contrast, Tokyo and New York have great mount of small GOS. In terms of large GOS, New York and London have adequate large GOS. While Shanghai and Tokyo are short of them. There is no significant difference in medium GOS. Meanwhile, based on the morphological features, we generalized four types of GOS which were normal GOS, riverside linear GOS and roadside linear GOS. For coastal GOS, only New York has great amount of them. Those of Shanghai and Tokyo are significantly fewer than those of New York. And there is no coastal GOS in London because the study area is not near the ocean. For the riverside linear GOS, except London which has only few amount, all the other three cities have certain amount of them. For roadside linear GOS, the four cities are different. Shanghai has masses of roadside GOS because of developing green belt encircling the city and the green way. Those of London are fewer than those of Shanghai and New York. Tokyo has hardly no roadside linear GOS.

5. DISCUSSION

5.1 Relations between availability and the distribution of GOS

From table 4 and table 6, we find that the high GOS coverage rate does not necessarily lead to a higher availability. London's coverage rate is lower than New York's, while it has a higher availability because of its great amount of large GOS. Besides, there are numerous small GOS in Tokyo and New York. On account of omitting the availability of GOS smaller than 0.25 ha, the availability of Tokyo and New York were probably underestimated. In addition, although the availability is not directly related to the population density, that the large population occupies copious urban space leads to a lower coverage rate. Hence the high-population-density cities have relatively lower availabilities.

5.2 Assessing the impacts of spatial pattern of GOS

In the aspect of small GOS, there are numerous this kind of GOS in Tokyo and New York. The small GOS has its unique characteristics. It has excellent convenience despite the shortage of functionality. For one thing, the small GOS can compensate for the deficiency of availability in part of the urban area; for another, it is pedestrian-friendly and convenient for the pedestrians to have a temporary rest. Thus, the small open space is a decent choice for the high-density cities Shanghai and Tokyo which do not have sufficient urban space. For the large GOS, Shanghai and Tokyo are obviously insufficient comparing with London and New York. Large GOS is applicable for diverse activities and can provide the visitors a wild perspective and rehabilitative experience (Lau et al., 2014; Giles-Corti et al., 2005). Shanghai and Tokyo are supposed to narrow the gaps with the two other cities.

The four cities are different in types of GOS. Firstly, the four cities have different amount of coastal GOS as a result of the limits of the study areas. The study area of London is not next to the sea and there is only a small part of the boarders of the study areas of Shanghai and Tokyo next to the sea. The coastal GOS will provide the visitors opportunities to enjoy the blue space which benefits the visitors' social activities and their mental health (De Bell et al., 2017; Völker et al., 2018). Shanghai and Tokyo can make a reasonable increase in the coastal GOS. Secondly, London has a disadvantage of riverside linear GOS. This type of GOS owns positive landscape values and can provide cultural services (Garcia et al., 2017; Vollmer et al., 2015). Thirdly, Shanghai and New York have more roadside linear open space than Tokyo and London. The roadside linear

https://doi.org/10.5194/isprs-archives-XLIV-M-2-2020-55-2020

GOS can enhance urban structural connectivity and improve citizen life satisfaction. In a word, the three types of GOS have diverse merits and are essential for an ecological city.

6. CONCLUSION

This study calculated the GOS indices and availability by using the spatial data of population and GOS of Shanghai, Tokyo, New York and London, and then analyzed the spatial patterns of GOS and its relation to availability. It was found that the four cities can be classified into two classes. Shanghai and Tokyo are high-population-density cities with medium GOS coverage and availability, and New York and London are medium-population-density cities with high GOS coverage and availability. Shanghai and London could increase the amount of small GOS. Because of developing green belt encircling the city and the green way, Shanghai has plenty of riverside and roadside linear GOS. London and Tokyo could consider adding linear GOS to improve the connectivity of GOS. In terms of future research directions, further study could explore the correlation between the spatial pattern of GOS and other urban issues such as ecosystem services and housing price.

REFERENCES

Abbasi, A., Alalouch, C., Bramley, G., 2016. Open Space Quality in Deprived Urban Areas: User Perspective and Use Pattern. *Procedia - Social and Behavioral Sciences* 216, 194-205.

Ambrey, C., Fleming, C., 2013. Public Greenspace and Life Satisfaction in Urban Australia. *Urban Studies* 51(6), 1290-1321.

Barbosa, O., Tratalos, J.A., Armsworth, P.R., Davies, R.G., Fuller, R.A., Johnson, P., Gaston, K.J., 2007. Who benefits from access to green space? A case study from Sheffield, UK. *Landscape and Urban Planning* 83(2-3), 187-195.

Cetin, M., 2015. Using GIS analysis to assess urban green space in terms of accessibility: case study in Kutahya. *International Journal of Sustainable Development & World Ecology*, 1-5.

Cho, S.-H., Lambert, D.M., Kim, S.G., Roberts, R.K., Park, W.M., 2010. Relationship between value of open space and distance from housing locations within a community. *Journal of Geographical Systems* 13(4), 393-414.

De Bell, S., Graham, H., Jarvis, S., White, P., 2017. The importance of nature in mediating social and psychological benefits associated with visits to freshwater blue space. *Landscape and Urban Planning* 167, 118-127.

Douglas, O., Russell, P., Scott, M., 2018. Positive perceptions of green and open space as predictors of neighbourhood quality of life: implications for urban planning across the city region. *Journal of Environmental Planning and Management* 62(4), 626-646.

Garcia, X., Benages-Albert, M., Pavón, D., Ribas, A., Garcia-Aymerich, J., Vall-Casas, P., 2017. Public participation GIS for assessing landscape values and improvement preferences in urban stream corridors. *Applied Geography* 87, 184-196.

Giles-Corti, B., Broomhall, M.H., Knuiman, M., Collins, C., Douglas, K., Ng, K., Lange, A., Donovan, R.J., 2005. Increasing walking: how important is distance to, attractiveness, and size of public open space? *American journal of preventive medicine* 28(2 Suppl 2), 169-176.

Hoffimann, E., Barros, H., Ribeiro, A.I., 2017. Socioeconomic Inequalities in Green Space Quality and Accessibility-Evidence from a Southern European City. *International journal of environmental research and public health* 14(8).

Ji, T., Pachi, A., 2005. Frequency and velocity of people walking. *Structural Engineer* 84(3), 36-40.

Kabisch, N., Strohbach, M., Haase, D., Kronenberg, J., 2016. Urban green space availability in European cities. *Ecological Indicators* 70, 586-596.

Kang, C.-D., 2015. The effects of spatial accessibility and centrality to land use on walking in Seoul, Korea. *Cities* 46, 94-103.

Kim, J., Nicholls, S., 2017. Using Geographically Weighted Regression to Explore the Equity of Public Open Space Distributions. *Journal of Leisure Research* 48(2), 105-133.

Lau, S.S.Y., Gou, Z., Liu, Y., 2014. Healthy campus by open space design: Approaches and guidelines. *Frontiers of Architectural Research* 3(4), 452-467.

Lovell, S.T., Taylor, J.R., 2013. Supplying urban ecosystem services through multifunctional green infrastructure in the United States. *Landscape Ecology* 28(8), 1447-1463.

Taylor, B.T., Fernando, P., Bauman, A.E., Williamson, A., Craig, J.C., Redman, S., 2011. Measuring the quality of public open space using Google Earth. *American journal of preventive medicine* 40(2), 105-112.

Völker, S., Heiler, A., Pollmann, T., Claßen, T., Hornberg, C., Kistemann, T., 2018. Do perceived walking distance to and use of urban blue spaces affect self-reported physical and mental health? *Urban Forestry & Urban Greening* 29, 1-9.

Vollmer, D., Prescott, M.F., Padawangi, R., Girot, C., Gret-Regamey, A., 2015. Understanding the value of urban riparian corridors: Considerations in planning for cultural services along an Indonesian river. *Landscape And Urban Planning* 138, 144-154.

Wolfe, M.K., Mennis, J., 2012. Does vegetation encourage or suppress urban crime? Evidence from Philadelphia, PA. *Landscape and Urban Planning* 108(2-4), 112-122.

Wu, C., Ye, X., Du, Q., Luo, P., 2017. Spatial effects of accessibility to parks on housing prices in Shenzhen, China. *Habitat International* 63, 45-54.

Wu, J., Wang, M., Li, W., Peng, J., Huang, L., 2015. Impact of Urban Green Space on Residential Housing Prices: Case Study in Shenzhen. *Journal of Urban Planning and Development* 141(4), 05014023.

Xu, C., Haase, D., Pauleit, S., 2018. The impact of different urban dynamics on green space availability: A multiple scenario modeling approach for the region of Munich, Germany. *Ecological Indicators* 93, 1-12.

This contribution has been peer-reviewed.
https://doi.org/10.5194/isprs-archives-XLIV-M-2-2020-55-2020 | © Authors 2020. CC BY 4.0 License.

The International Archives of the Photogrammetry, Remote Sensing and Spatial Information Sciences, Volume XLIV-M-2-2020, 2020
ASPRS 2020 Annual Conference Virtual Technical Program, 22–26 June 2020

DIGITAL SURFACE MODEL DERIVED FROM UAS IMAGERY ASSESSMENT USING HIGH-PRECISION AERIAL LIDAR AS REFERENCE SURFACE

J. Lopez[1], R. Munjy[1]

[1] Department of Civil and Geomatics Engineering, Lyles College of Engineering, California State University, Fresno, 2320 E San Ramon Ave., Fresno, CA 93740, USA – (jakelopez1@mail.fresnostate.edu, riadhm@csufresno.edu)

KEY WORDS: Point Cloud, Aerial LiDAR, Unmanned Aerial System, Orthomosaic, DEM, Photogrammetry, GNSS

ABSTRACT:

Imagery captured from aerial unmanned systems (UAS) has found significant utility in the field of surveying and mapping as the efforts of the computer vision field combined the principles of photogrammetry. Its respectability in the remote sensing community as increased as the miniaturization of on-board survey-grade global navigation satellite system (GNSS) signal receivers has made it possible to produce high network accuracy contributing to effective aerotriangulation. UAS photogrammetry has gained much popularity because of its effectiveness, efficiency, economy, and especially its availability and ease of use. Although photogrammetry has proven to meet and exceed planimetric precision and accuracy, variables tend to cause deficiencies in the achievement of accuracy in the vertical plane. This research aims to demonstrate achievable overall accuracy of surface modelling through minimization of systematic errors at a significant level using a fixed-wing platform designed for high-accuracy surveying with the eBee Plus and X models by SenseFly equipped with survey-grade GNSS signal-receiving capabilities and 20MP integrated, fixed-focal length camera. The UAS campaign was flown over a site 320 m by 320 m with 81 surveyed 3D ground control points, where horizontal positions were surveyed to 1.0 cm horizontal accuracy and 0.5 cm vertical accuracy using static GNSS methods and digital leveling respectively. All AT accuracy was based on 75 independent checkpoints. The digital surface model (DSM) was compared to a reference DSM generated from high-precision manned aerial LiDAR using the Optech Galaxy scanner. Overall accuracy was in the sub-decimeter level vertically in both commercial software used, including Pix4Dmapper and Agisoft Metashape.

1. INTRODUCTION

Unmanned aerial systems equipped (UAS) equipped with cameras are able to create 3D point clouds for precise modeling for large area mapping and reconstruction using the principles of photogrammetry. In practice, aerial photogrammetry remains an established and effective method to for collecting accurate and precise planimetric and elevation data. Aerial LiDAR has become another method of effectively collecting elevation data and has some advantages over aerial photogrammetry including the ability to collect data outside of optimal sunlight conditions and multiple pulse returns allowing penetration through vegetation to capture ground points. Barring the ability to penetrate vegetation to capture ground and fly outside of optimal sunlight and sun angle, obtaining network-accurate and precise elevation data through use of UAS imagery capture is a highly, efficient, and economic method. The purpose of this study is to compare, and therefore validate, the collection of elevation data using 20MP resolution imagery captured from UAS equipped with on-board, survey-grade global navigation satellite system (GNSS) capabilities. The elevation data will be assessed in the form of digital surface model (DSM) of unobstructed, solid surfaces including paved asphalt road and rooftops generated from imagery-based point cloud computed through two popular commercial structure-from-motion software for processing UAS imagery, Pix4Dmapper and Agisoft Metashape.

There is a high saturation of studies throughout the literature demonstrating proof-of-concept of using UAS imagery for high precision mapping using surveyed ground control for network orientation (Agüera-Vega, Carvajal-Ramírez, & Martínez-Carricondo, 2017; Eisenbeiß, 2009; Eisenbeiss & Sauerbier, 2011). Consequently, point clouds and surface models have been investigated throughout the literature. However, less work has been done comparing the elevation products of point clouds and surface models produced from UAS imagery to LiDAR. There is a gap in the literature comparing DSM from UAS imagery to a DSM generated from high-precision aerial LiDAR. This is likely due to the large cost associated with planning a full-scale, manned aerial LiDAR mission. In addition, there is lack of research including a significantly large sample checkpoints as well as of flights and flight orientations using a calibration site with a highly densified control network for the assessment of digital surface models on hardscaped areas such as roofs and asphalt roads.

Various research has assessed imagery-derived point clouds and DSM's. Research has been performed to investigate the effect of flight planning, including flight orientation, on the quality of the point cloud from a sample of only 6 check points, holding 20 points as ground control (Chiabrando, Lingua, Maschio, & Losè, 2017). While the AT results were at the centimeter-level, the sample size was insufficient to ensure repeatability to achieve that level of accuracy. The quality of the point clouds were assessed from capture at different camera orientations and the point density of vertical cross sections was used as a measure of quality rather than comparison to reference surface. A study assessed imagery-derived point cloud from UAS that analyzed the effect of automatic and semi-automatic GCP measurement on the final georeferenced cloud using checkpoints surveyed by total station (Harwin & Lucieer, 2012). Anders and colleagues performed a number of flight campaigns to assess the accuracy of surface modelling finding an average vertical discrepancy in the decimeter range (Anders, Masselink, Keesstra, & Suomalainen, 2013). Although 50 GCP were used, it was unclear how they were used in the final computation of report of vertical discrepancies.

This contribution has been peer-reviewed.
https://doi.org/10.5194/isprs-archives-XLIV-M-2-2020-61-2020 | © Authors 2020. CC BY 4.0 License.

Although point clouds have been thoroughly assessed through means of checkpoint validation, elevation models derived from UAS imagery have less been compared to reference surfaces generated from LiDAR methods. A study compared UAS imagery-derived DSM to terrestrial laser scan (TLS)-derived DSM, as a reference surface, of a man-made dike and processing the imagery in Pix4Dmapper and Agisoft Photoscan (now Metashape) (Naumann, Geist, Bill, Niemeyer, & Grenzdörffer, 2013). The comparison in this case applied a simple DSM subtraction and an elevation heat map across the entire surface as well as various segment surfaces was analyzed. Although results were based on a large sample, there was no mention of checkpoint statistics nor of specific distribution of the 11 surveyed ground control point (GCP) distribution. Przybilla and colleagues also compared UAS imagery-based point cloud to TLS point cloud along cross-section vertical profiles (Przybilla, Lindstaedt, & Kersten, 2019). The author used Pix4Dmapper and Agisoft Metashape to process the imagery into point clouds. An aspect of this study was to investigate an effect of image format (TIF or JPG) on point cloud accuracy as well as an effect of image down sampling for keypoint extraction. The problem with this study is that the checkpoints for the point cloud were used as the control points in the AT process, therefore the results of this study are questionable. Kosmatin Fras and colleagues assessed the accuracy of UAS imagery-derived DSM for about 20 checkpoints without the use of a reference surface such as that derived from either terrestrial or aerial LiDAR (Fras, Kerin, Mesarič, Peterman, & Grigillo, 2016).

The study of this paper is an extension of an earlier study over the same site that compared UAS imagery-derived point clouds to a reference TLS point cloud (Peterson, Lopez, & Munjy, 2019). In this study, imagery was collected using a senseFly eBee Plus model equipped with survey-grade on-board GNSS collection capabilities. From these data a sparse point cloud was adjusted by AT and validated to be a 2-3cm horizontally accurate and 4 cm vertically accurate according to statistics based on 75 checkpoints surveyed by static-GNSS methods with elevations resolved by digital leveling. A vertical profile 1 cm-thick was extracted from both UAS imagery and TLS point clouds along 200 m of asphalt road and compared. An unresolved vertical bias of nearly 15 cm was found. Planar regression was applied to point samples on roof faces for both point clouds and compared where the vertical bias was not present such as was found along the asphalt road.

In the study of this paper, a high-precision, manned aerial LiDAR mission was performed over a site located in the foothills of the Sierra Nevada Mountain Range located in Fresno County for use as reference data for comparison. The site was densified by 81 high accuracy ground control points with horizontal positions established to 1 cm using static GNSS method and vertical positions to 0.5 cm using digital leveling. AT results were optimally established in the sparse point cloud before being densified and converted to DSM for surface comparison. Orthophoto generation was performed for the purpose of checkpoint location in the DSM for checkpoint validation of the surface. Cross-sectional profiles were extracted from all DSM's, including the reference DSM from aerial LiDAR, along asphalt roads bisecting the site and along rooftops. Along with these comparisons, results were compiled and tabulated for flying height and flight line orientation.

2. METHODOLOGY

2.1 Site

A calibration site 320m by 320m located about 32km north of the campus of California State University, Fresno in the foothills of the Sierra Nevada mountain range was designated for this study. The site has rolling hills topography with a control point elevation range of about 20m.

2.2 Control

Control points (80) were surveyed horizontally using static GNSS methods, observed with simultaneous setups for 8 hours on multiple occasions producing a final accuracy of 1.0cm after adjustment. The vertical positions were resolved by differential leveling producing a final accuracy of 0.3cm. A total of 80 control points were spaced approximately 40m in a 9-by-9 square grid distributed throughout the site (Figure 1). The ground targets were designed black and white, circular, 42cm-diameter (Figure 2).

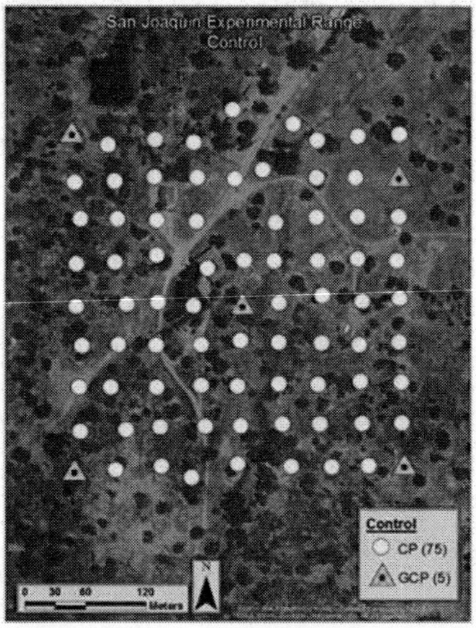

Figure 1. Research site at San Joaquin Experimental Range displayed with the total layout of 80 control points including GCP and CP configuration.

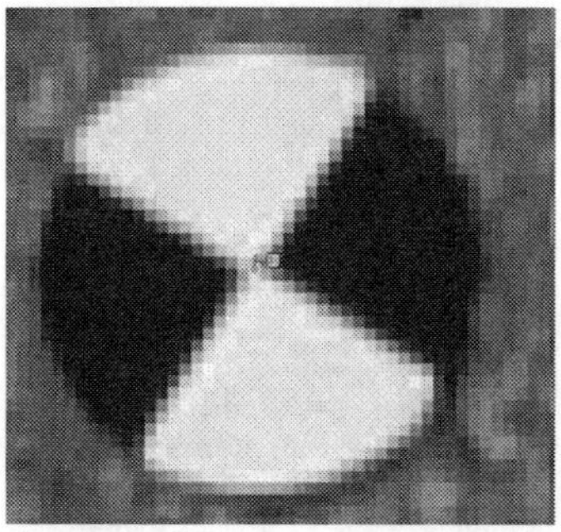

Figure 2. Ground control target as it appears in photograph.

2.3 Aerial LiDAR Flight Campaign

A high precision aerial LiDAR scan (ALS) flight was performed over the site by manned-helicopter using an Optech Galaxy scanner. The ALS elevation product was used as a reference surface for comparison. Table 1 summarizes the flight parameters and network accuracy of this flight.

Sensor	Optech Galaxy Prime
Flying Height Above Terrain (m)	200
Pulse Rate (Khz)	300
RMSEz (cm)	1
RMSExy (cm)	3-5
Point Density (pulses/m^2)	127
Average Point spacing (cm)	11
Scan Angle (deg)	43
LASER Foot print at Nader (cm)	5
LASER Foot print at Edge (cm)	9

Table 1. High precision aerial LiDAR campaign parameters used as reference surface.

2.4 UAS Hardware

Two models of fixed-wing aircraft, the SenseFly eBee Plus (Plus) and eBee X (X), were used for this research. Both UAS platform models were equipped with survey-grade, dual-frequency (L1, L2) GNSS receiver antennas and are able to resolve positions by the method of post-processing-kinematic (PPK). The optical sensor integrated into both eBee models (S.O.D.A.). The optical sensor integrated into the X model has oblique-angle image capture capability, which was not used in this research. The S.O.D.A. camera has a global mechanical shutter, CMOS sensor, 20MP sensor resolution, 2.4µm pixel size, and a published focal length of 10.6mm. More specifications of the camera are given in the Table 2.

Imaging sensor	S.O.D.A.
Sensor resolution	20 MP
CCD dimensions	3:2 / 5472 x 3648
Pixel size	2.4 microns
Maunfactured focal length	10.6 mm
Sensor type	BSI CMOS 1-inch
Shutter	Global mechanical
Maximum shutter speed	1/2000s
ISO	125-6400
Aperture	F2.8-11
Sensitivity	450/520/660 nm
White balance	Automatic
Output format	JPEG and DNG
Dust & shock protection	Built-in

Table 2. Sensefly eBee UAS platform hardware specifications.

2.5 UAS Flight Campaigns

A total of 3 flights were performed on separate dates. Two flights were performed with the Plus model at different flying heights, and one flight with the X model. Table 3 summarizes the flight plan parameters including the number of lifts required by the platform to cover the site, the aircraft model, flying height above ground level (AGL), ground sample distance (GSD), the cardinal direction, number of photos, average overlap (O) and side overlap (S), and wind speed and direction. In addition, the flight code is used to distinguish the flights by flying height AGL and flight path direction.

Lifts	Aircraft	Flying Height (m AGL)	GSD (cm)	Direction	Photos	Avg O/S (%)	Wind (km/h)	Wind Dir (Deg)	Code
2	eBee Plus	114	2.45	EW-NS	586	60/80	15	1.1	A
2	eBee Plus	84	1.91	EW-NS	953	60/80	12	348.8	B
1	eBeeX	105	2.37	NS	709	78/78	11	320.1	C

Table 2. Summary of UAS flight campaign plans.

The Plus was flown on two separate dates (Flights A, B) and the X once on another date (Flight C). Each of the Plus flights were performed 2 times, once in an east-to-west (EW) and another time in a north-to-south (NS) flight-line orientation. A third flight line configuration was studied where both orientations were combined to form perpendicular flight-lines.

2.6 Software

Agisoft Metashape and Pix4Dmapper were used in this research. Both software were among the most popular commercial sfm software used for processing UAS imagery at the time of this study.

2.7 Software Parameters

Images were matched and aligned with image sizes at full scale resolution. Aerotriangulation was performed using a constrained solution with 5 GCP and accuracy was assessed using the remaining 75 control points as checkpoints (CP) prior to point cloud generation. The 5 GCP layout is given in Figure 1. Exterior orientation positions were initially assigned *a priori* uncertainty values of 2 to 3cm horizontally and vertically according to system accuracy. An adjustment was performed and uncertainty values were edited to maintain residual errors within two times the uncertainty. Self-calibration was performed during adjustment similarly in both software adjusting focal length (*f*), principal point (c_x, c_y), three coefficients for symmetric radial distortion (*k1, k2, k3*), and two coefficients of tangential distortion (*p1, p2*). The AT results include only ground control points, therefore the residuals were analyzed to ensure that the adjustment was not over- or under-constrained.

2.8 Analysis

An unclassified point cloud was generated from the 5 GCP AT results. All checkpoints were omitted from the software project prior to point cloud generation to eliminate the uncertainty of any proprietary software influence on checkpoint residuals. The point cloud was generated with a density of 400 pls/m^2 and converted to a 5cm digital surface model (DSM). A 2.5cm orthomosaic was created and 2D CP coordinates were measured using ESRI ArcMap. Elevations at the 2D CP locations were interpolated in both UAS imagery and ALS DSM's using ArcMap then compared to the surveyed ground control and to each other. Several elevation path profiles were extracted along paths over roofs and roads, including a driveway, from the ALS and UAS DSM's and compared (Figure 3).

Figure 3. Features and profile paths for elevation profile analysis.

Flight data were separated and analyzed at checkpoints according to flight path direction including east-west (EW), north-south (NS), and perpendicular, or cross, flight paths (EW-NS). All flight-line-oriented configurations were performed two times, once in Agisoft Metashape and once in Pix4Dmapper resulting in a sample of 14 point cloud and orthomosaic results, i.e. 7 point clouds per software. The sample of results were assessed using the root-mean-square-error statistic horizontally (RMSExy) for the orthomosaic and vertically (RMSEz) for the point cloud/DSM. However, elevation profile accuracy assessment did not include investigation into the effect of flight path direction.

3. RESULTS

3.1 DEM Analysis at Checkpoints

3.1.1 Pix4D Pix4D orthomosaic accuracy did not vary substantially by flying height nor by direction (average RMSExy = 0.9 ± 0.1cm). The differences between the maximum and minimum error shown by the range statistic was least for flight B. The average vertical error range was higher in magnitude and uncertainty (8.5 ± 3.4cm) than the average horizontal error range (4.0 ± 0.8cm). An overall average negative error vertically indicated that the adjustment was biased in the negative direction. This was true for the higher altitude but generally negative for all flights (-1.2 ± 0.7cm) (Table 3).

Orientation	Flight	RMSEz	RMSExy	Std_dz	Std_dxy	Avg_dz	Avg_dxy	Range_dz	Range_dxy
EW	A1	2.1	0.9	1.4	0.8	-1.6	1.5	7.6	3.5
NS	A2	2.7	1.0	1.6	1.1	-2.1	1.9	9.4	5.2
EW-NS	A3	2.2	0.8	1.5	0.9	-1.6	1.3	10.8	4.0
EW	B1	1.5	0.9	1.3	1.0	-0.8	1.7	5.6	4.0
NS	B2	1.6	0.8	1.3	0.7	-0.9	1.3	5.9	3.7
EW-NS	B3	1.6	0.8	1.2	0.7	-1.2	1.3	5.5	2.9
NS	C	2.5	1.0	2.5	1.0	0.2	1.8	14.7	4.6
Average		**2.0**	**0.9**	**1.5**	**0.9**	**-1.2**	**1.6**	**8.5**	**4.0**
St. Dev		**0.5**	**0.1**	**0.5**	**0.1**	**0.3**	**0.1**	**3.4**	**0.8**

Table 3. Results of DEM analysis at 75 checkpoints for Pix4D.

This contribution has been peer-reviewed.
https://doi.org/10.5194/isprs-archives-XLIV-M-2-2020-61-2020 | © Authors 2020. CC BY 4.0 License.

3.1.2 Metashape In flights A and B, the parallel flight paths (NS A1, A2, EW B1, B2) resulted in vertical accuracy that was about 2 times better than the results of the orthogonal flight paths (e.g. RMSEz A1 = 1.3cm, A3 = 3.6cm, B1 = 1.7cm, B3 = 4.7cm). Horizontal accuracy was consistently slightly above 1.0cm (Average RMSEz = 1.1 ± 0.1cm) for all flights indicating that the effect of varying flight parameters was negligible. The average horizontal error range (6.5 ± 1.2cm) was greater than in Pix4D (4.0 ± 0.8cm), which showed a more uniform adjustment throughout the control region in this software (Table 4).

Orientation	Flight	RMSEz	RMSExy	Std_dz	Std_dxy	Avg_dz	Avg_dxy	Range_dz	Range_dxy
EW	A1	1.3	1.1	1.1	1.0	-0.7	2.4	5.0	4.0
NS	A2	1.7	1.1	1.4	1.3	-1.0	2.6	6.6	5.4
EW-NS	A3	3.6	1.1	1.2	1.2	-3.4	2.6	5.6	5.9
EW	B1	1.7	1.0	1.6	1.1	0.8	2.0	7.6	5.0
NS	B2	2.0	1.2	1.2	1.1	1.6	2.7	6.4	6.4
EW-NS	B3	4.7	1.2	1.3	1.2	-4.5	2.9	6.2	5.4
NS	C	2.8	1.2	2.0	1.3	-2.0	2.7	8.4	4.9
Average		2.6	1.1	1.4	1.2	-1.3	2.6	6.5	5.3
St. Dev		1.2	0.1	0.3	0.1	0.8	0.1	1.2	0.7

Table 4. Results of DEM analysis at 75 checkpoints for Metashape.

3.1.3 Reference Surface: The vertical accuracy of the high precision ALS point cloud was 1.5cm (1.5 ± 1.0cm) based on a 74 checkpoint sample. The negative average vertical error (-1.1cm) indicated a negative bias. The vertical error range was 5.7cm, making it comparable to the noise exhibited by UAS point cloud (Table 5).

CP	74
Statistic	cm
RMSEz	1.5
Std_dz	1.0
AbsAvg_dz	1.3
Avg_dz	-1.1
Median_dz	-1.1
Max_dz	1.8
Min_dz	-3.9
Range_dz	5.7

Table 5. DEM analysis of reference surface using data collected from Optech Galaxy LiDAR mission.

3.2 Elevation Profile Analysis

3.2.1 Pix4D: Average error was for all profiles (n = 140) from DSM's generated by Pix4Dmapper including various roofs and roads was -1.0cm with a standard error of the mean of 0.1cm. The standard error of the mean was computed by dividing the standard deviation by the square root of the sample. Overall accuracy for surfaces resulted in an average RMSE of 2.0 ± 0.7cm. Table 6 below provides a summary of the profile statistics.

n = 140	Avg	Min	Max	St.dev.	RMSEz
Average	-1.0	-3.5	3.0	1.2	2.0
St.dev.	0.1	1.6	2.2	0.4	0.7

Table 6. Summary of elevation profile differences between reference surface and surface generated by UAS imagery for Pix4D

3.2.2 Metashape: The average error of all elevation profiles from DSM's generated by Agisoft Metashape was -0.7cm with a standard error of the mean of 0.2cm using formula 1 above. Average RMSEz was slightly greater than Pix4D at 2.2 ± 1.2cm with a lesser precision. Table 7 below provides a summary of statistics.

n = 140	Avg	Min	Max	St.dev.	RMSEz
Average	-0.7	-3.1	3.2	1.1	2.2
Stdev	0.2	2.3	2.7	0.4	1.2

Table 7. Summary of elevation profile differences for Metashape

https://doi.org/10.5194/isprs-archives-XLIV-M-2-2020-61-2020

The range of error was similar to Pix4D with an average error range of approximately 6 to 8cm. Error range was large along the road (Figure 4). Overall, all flight elevation profile results agreed with the results at checkpoints with a vertical RMSE around 2cm.

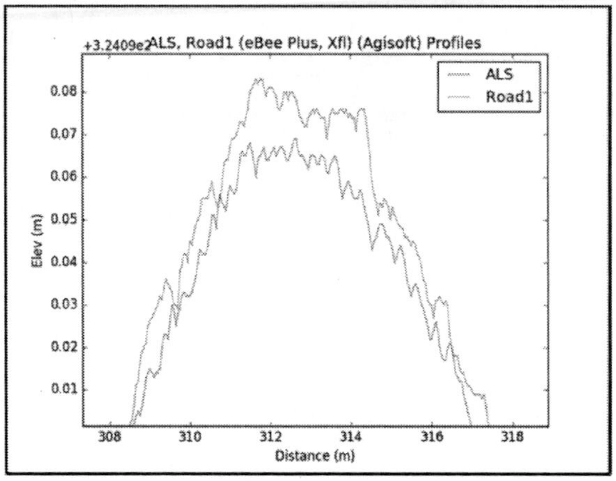

Figure 4. Sample of graphical display of elevation profile differences behavior along road 1 (see Figure 3 for road 1).

In Figure 4 the vertical bias between the systems was also apparent. Errors of greatest magnitude occurred at roof peaks. An example of this occurrence is given in Figure 5.

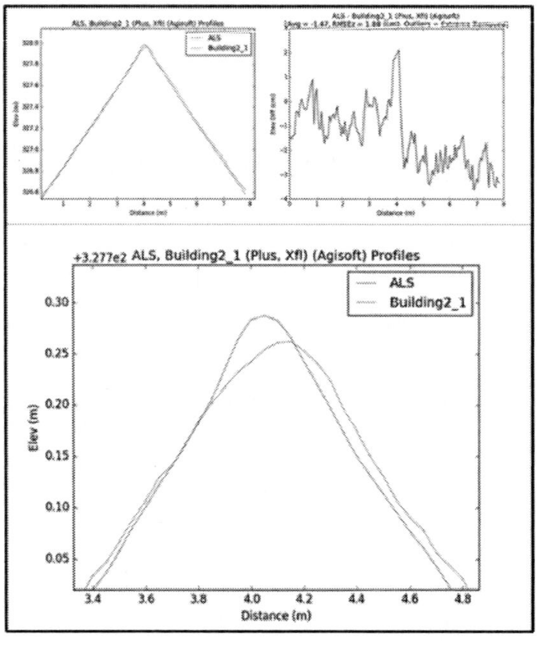

Figure 5. Sample graphical display of vertical discrepancy due to resolution of roof peak feature along building 2 (see Figure 3)

The graph of differences (top right graph of Figure 5) showed that the errors of greatest magnitude occurred at the peak and errors changed direction beyond that. In the larger scale profile graph (bottom) To the right of the peak a vertical separation occurred along the remainder of the roof side.

4. DISCUSSION

In this discussion it will be important to recall the accuracy of the surveyed ground control for ASPRS accuracy classification. This research will be classified according to product accuracy per ASPRS positional accuracy standards of 2014 (Table 8) since horizontal accuracy was determined by the orthomosaic and vertical accuracy by the point cloud-derived DSM. Product accuracy will be assessed with the ground control as 1.0cm since the vertical accuracy of the surveyed ground control less than 0.5cm and horizontal 1.0cm.

Aerial Triangulation and Ground Control Accuracy Requirements, Orthoimagery and/or Planimetric Data Only And Elevation Data					
Product Accuracy (RMSEx, RMSEy) (cm)	Product Accuracy (RMSEz) (cm)	A/T Accuracy		Ground Control Accuracy	
		RMSEx and RMSEy (cm)	RMSEz (cm)	RMSEx and RMSEy (cm)	RMSEz (cm)
7	4	4	2	2	1
18	4	9	2	4	1

Table 8. Summary of results based on ASPRS 2014 standards for positional accuracy standards.

In the checkpoint results from Pix4D the vertical accuracy may have been affected by flying height but the average RMSEz was 2.0cm ± 0.5cm. The increased altitude of flight A (114m), compared to flights B (84m) and C (105m), displayed lower accuracy. The effect of flying height was seen in accuracy from flights B (84m AGL) to flight C (104m AGL). The flight-line directions did not have an effect in the Pix4D results between flights A and B for each flight. The accuracy also may have been a function of the GSD.

Orthogonal flight paths processed in Metashape displayed a degradation of vertical accuracy in the checkpoints when compared to the results of the parallel flight lines. The effect of orthogonal flight lines were seen in the overall standard deviation of the accuracy (Average RMSEz = 2.6 ± 1.2cm) at checkpoints.

The magnitude of errors that occurred at roof peaks that was observed in all flights was a possible function of the incidence angle. Further, to the right of the peak in the profile graphs a

https://doi.org/10.5194/isprs-archives-XLIV-M-2-2020-61-2020 | © Authors 2020. CC BY 4.0 License.

vertical separation was found indicating that a bias may have been caused by angle of incidence. A greater degree of spatial frequency along the road and asphalt quality may have caused the DSM to be noisier than roof profiles (Figure YY).

5. DISCUSSION

This research demonstrated achievable accuracy of UAS imagery-based horizontal and elevation products using SenseFly's fixed-wing, dual-frequency GNSS-capable eBeeX and eBee Plus models (3 flights total). A total of three flight campaigns were performed on separate dates over the San Joaquin Experimental Range located in Fresno County, California in the foothills of the Sierra Nevada Mountain Range over a site with 81 surveyed control point flight targets. From the UAS-captured imagery point clouds, surface models, and orthomosaics were generated using Pix4Dmapper and Agisoft Metashape, two popular, commercial SfM software used for processing UAS imagery. A reference surface was generated from a high-precision Optech Galaxy LiDAR scan point cloud from manned-helicopter. The orthomosaic was assessed for horizontal accuracy and DSM for vertical accuracy at 75 checkpoints. Nearest neighbor interpolation was performed at vertical checkpoints using ArcMap. Various elevation profiles were extracted from UAS and ALS DSM's along roofs and roads and compared for assessment of accuracy to observe elevation accuracy between checkpoints.

ACKNOWLEDGEMENTS

Special thanks and acknowledgement to California State University Education Foundation, John Erickson and the office of Photogrammetry and Preliminary Investigations at California Department of Transportation for supporting this research though funding and direction. Thanks to companies Woolpert, Jones Synder & Associates, and Towill for their contribution in conducting the UAS campaigns. Thanks to Towill Inc. for graciously offering to support this research in flying the aerial LiDAR mission because without them there would be no reference surface for comparison. Thanks to Dr. Yushin Ahn at CSU, Fresno for his help with this paper. Thanks to Dr. Mike Mustafa Berber and Dr. Scott Peterson for their support and painstaking effort in surveying the ground control at the research site.

REFERENCES

Agüera-Vega, F., Carvajal-Ramírez, F., & Martínez-Carricondo, P., 2017. Assessment of photogrammetric mapping accuracy based on variation ground control points number using unmanned aerial vehicle. *Measurement, 98*, 221-227.

Anders, N., Masselink, R., Keesstra, S., & Suomalainen, J. (2013). High-res digital surface modeling using fixed-wing UAV-based photogrammetry. *Proceedings of the Geomorphometry, Nanjing, China*, 16-20.

Chiabrando, F., Lingua, A., Maschio, P., & Losè, L. T. (2017). The influence of flight planning and camera orientation in UAVs photogrammetry. A test in the area of Rocca San Silvestro (LI), TUSCANY. *The International Archives of Photogrammetry, Remote Sensing and Spatial Information Sciences, 42*, 163.

Eisenbeiß, H. (2009). *UAV photogrammetry.* ETH Zurich,

Eisenbeiss, H., & Sauerbier, M. (2011). Investigation of UAV systems and flight modes for photogrammetric applications. *The Photogrammetric Record, 26*(136), 400-421.

Fras, M. K., Kerin, A., Mesarič, M., Peterman, V., & Grigillo, D. (2016). ASSESSMENT OF THE QUALITY OF DIGITAL TERRAIN MODEL PRODUCED FROM UNMANNED AERIAL SYSTEM IMAGERY. *International Archives of the Photogrammetry, Remote Sensing & Spatial Information Sciences, 41.*

Harwin, S., & Lucieer, A. (2012). Assessing the accuracy of georeferenced point clouds produced via multi-view stereopsis from unmanned aerial vehicle (UAV) imagery. *Remote Sensing, 4*(6), 1573-1599.

Naumann, M., Geist, M., Bill, R., Niemeyer, F., & Grenzdörffer, G. (2013). Accuracy comparison of digital surface models created by unmanned aerial systems imagery and terrestrial laser scanner. *International Archives of the Photogrammetry, Remote Sensing and Spatial Information Sciences, 1*, W2.

Peterson, S., Lopez, J., & Munjy, R. (2019). COMPARISON OF UAV IMAGERY-DERIVED POINT CLOUD TO TERRESTRIAL LASER SCANNER POINT CLOUD. *ISPRS Annals of Photogrammetry, Remote Sensing & Spatial Information Sciences, 4.*

Przybilla, H.-J., Lindstaedt, M., & Kersten, T. (2019). INVESTIGATIONS INTO THE QUALITY OF IMAGE-BASED POINT CLOUDS FROM UAV IMAGERY. *International Archives of the Photogrammetry, Remote Sensing & Spatial Information Sciences.*

DEVELOPING GEOSPATIAL SCIENTISTS – USING STUDENTS AS PARTNERS IN DRONE RESEARCH

T. McKee, L. Hashemi-Beni

Department of Geomatics and Built Environment, North Carolina Agricultural and Technical State University, Greensboro, North Carolina - tlmckee@aggies.ncat.edu - lhashemibeni@ncat.edu

KEY WORDS: Geospatial, Students, Research, Data, GIS, Drone, Education, STEM

ABSTRACT:

The U.S. Department of Labor Occupational Outlook Handbook predicts geospatial careers to increase much faster than average (15%) from 2018 to 2028 with no additional on-the-job training expected. Geospatial professionals can assist in promoting these career opportunities by mentoring high school students through real-world and transferable-skill building activities. The aim is to enhance students' awareness and stimulate their interest towards STEM education and careers, especially in geospatial data analytics. This area of study incorporates a variety of modern-day tools for analyzing and mapping the Earth. The technology used offers a radically different way in which geospatial scientist produce and use the geospatial information required to manage a large variety of communities and industries.

1. INTRODUCTION

Geospatial Science is an area of study that incorporates a variety of modern-day tools for analyzing and mapping the Earth (American Association for the Advancement of Science, 2020). It includes the development and applications of traditional geospatial technologies, such as geographic information systems (GISs), global positioning systems (GPSs) and remote sensing (RS) (National Research Council, 2013) as well as continuously evolving and interdisciplinary technologies such as drones, miniature sensors, real-time spatiotemporal data analytics, crowdsourcing, geolocation, and wireless infrastructure (Federal Geographic Data Committee, 2016).

These technologies individually present dynamic challenges. The workforce required to administer, deploy and display our technologies cannot be uniformly trained (Hammond, et al., 2019). Traditional K-12 learning environments would be ideal but these environments are not designed to specialize in geospatial learning because the educational landscape varies across US school systems, such as how much geospatial content is taught, how the content is structured, and how the content emphasizes scientific careers (Petcovic, et al., 2018). Additionally, it is estimated that only 3% of secondary educators are formally trained in a geospatial field of study (American Geoscience Institutue , 2010). Therefore, geospatial scientists must be engaged in mentoring and performing discipline specific educational research to determine best practices for training future geospatial scientists on these continuously evolving technologies. Accordingly, the following research aims to orient high school students to geographic information system mapping and drone research using an applied data literacy approach.

For the purpose of this research, data literacy is defined as the ability to plan for, document, analyze and visualize data. Geospatial curriculum tasks will be guided by the U.S. Department of Labor, Employment and Training Administration's Geospatial Technology Competency Model (GTCM) (U.S. Department of Labor, Employment and Training Administration, 2020). The GTCM is a 9-tier pyramid that illustrates industry, academic, workplace and personal competencies that geospatial professionals should seek to acquire. The specific knowledge, skills, and abilities recommended for success are listed for each individual competency. The GTCM is designed to assist users with a variety of objectives, such as geospatial curriculum development, geospatial workforce needs and career readiness goals. Additionally, the model provides useful tools, such as a pre- and post- geospatial skills gap assessment to assist with curriculum or training alignment.

K-12 data literacy approaches are also supported by the National Academies of Sciences, Engineering and Medicine in a recent report: *Science and Engineering for Grades 6-12: Investigation and Design at the Center* (2019), which states that students in grades 6-12 gain knowledge and reasoning abilities by manipulating and visualizing data they have collected themselves versus data obtained from secondary sources (National Academies of Sciences, Engineering, and Medicine, 2019). They acknowledge that the ability to work with data is at the core of science and engineering. Therefore, students should have in-depth opportunities to develop a data skillset. This issue is critical for STEM students because prior learning experiences shape student identities, knowledge, attitude, and understandings in future learning environments (Roberts, et al., 2018). Out-of-school programs and informal experiences that engage and train students using real-world data drawn from their local environment connect students to their communities and may assist in increasing the number of geospatial scientists in the US workforce.

The remainder of the article is outlined in the following sections: Developing Geospatial Scientists, Theoretical Framework, Methodology, Geospatial Summer Program Characteristics and Future Recommendations.

This contribution has been peer-reviewed.
https://doi.org/10.5194/isprs-archives-XLIV-M-2-2020-69-2020 |

1.1 Developing Geospatial Scientists

What is often not considered by geospatial professionals is who among us will be responsible for mentoring and training future geospatial scientists using authentic real-world research experiences that incorporate relevant geospatial technologies (Huntoon, et al., 2005). It is a fallacy to make an assumption that K-12 science educators alone will do this; they face many challenges within their own field of study.

In general, science educators in K-12 environments face a range of administrative challenges; from how a class is evaluated to the time allotted for instruction (Ayotte-Beaudet, et al., 2017). Science educators are further limited by factors, such as the assumption that geospatial concepts are of less rigor, lack of access to geospatial technology, training, and lack of financial support for geospatial programs (Orion, 2019). Osborne et al. (2020) conducted research that suggested even geospatial educators may lack interest or even knowledge of how geospatial technology is related to their own high school geography curriculum (Osborne, et al., 2020). This lack of educator exposure to geospatial technology, workshops, and coursework can negatively influence 6th thru 12th grade student career aspirations. For example, the number of undergraduate students pursuing geoscience degrees has not significantly changed since 2012 (Keane, 2018). Additionally, 50% of students entering STEM fields at the undergraduate level do not finish (Seymour & Hunter, 2019).

The preparation of students during 6th through 12th grade is a contributing factor for success. Students must plan and establish an academic schedule that permits them to transition towards science careers prior to their 11th or 12th grade of high school because K-12 science tracks are often designed to begin in 9th grade. Hands-on opportunities and geospatial engagement at local levels will be key to bridging this academic disconnect.

1.2 Informal Learning Environments

With so many limitations on traditional K-12 education, informal learning environments represent an optimum avenue to expose students to geospatial research and activities. This approach is critical because research has shown that students exposed to careers at an early age are more aware of what skills they need to develop to enter in to a said career (Cagle, et al., 2018). Unfortunately, geospatial careers are often not presented as an option

To promote the progress of geospatial science, geospatial scientists must be committed to engaging in participatory research with students with an understanding that the next generation of geoscientists are being mentored and trained (Hawthorne & Jarrett, 2018). Students engaged in mentored experiences are free to offer new perspectives on geospatial research. Additionally, Historically Black Colleges and Universities (HBCUs) are traditionally positioned to be key partners for recruiting, training and mentoring underrepresented geoscientists; especially within K-12 academic levels and informal environments (Archer, et al., 2018).

2. THEORETICAL FRAMEWORK

The theoretical framework for the study is based on how students learn in informal environments (National Research Council, 2009). Students are not simply learning conceptual knowledge and scientific skills; they are learning the culture of science, understanding how it blends with their daily life, how it is relevant to them and how it requires interactive collaboration (Fenichel & Schweingruber, 2010). Informal learning becomes an extension of their environment and a source of personal enjoyment (i.e. watching TV, participating in a program, volunteering, visiting museums).

3. METHODOLOGY

Program implementation was initially planned as a one-week day summer program with high school students on a university campus. However, due to COVID-19, the program is transitioning to an online learning environment. Therefore, the program itself will be discussed briefly.

3.1 Pre-Program Tasks

Students were scheduled to complete pre-program tasks designed to provide research context and background information Examples include but are not limited to online questionnaires, videos, getting-to-know each other activities and program registration.

3.2 Recruitment

Eighteen high school students from Guildford County, North Carolina were recruited to participate in the GIS & Drones Program. Recruitment efforts were assisted through previously established university and K-12 community partnerships.

3.3 Program Implementation

Program implementation will focus on two major themes: 1) GIS Mapping and 2) Drone Data Processing and Visualization. Students will build GIS and Drone research-based portfolios. The portfolios serve as artifacts of what students have learned and accomplished. Each student will select up to two (2) portfolio documents to present at the end of the program. Students are required to include a minimum of six documents in their final portfolio. Each document aligns with and reflects one major task.

3.4 Evaluation

Evaluation will occur across three (3) levels: students, programs, and parents. A logic model will guide the evaluation (Montano, 2016). Evaluations will consist of pre and post program surveys, focus group discussions, instructor observations, and student portfolios. A pilot study of all surveys will be conducted to make necessary revisions to the interview instruments and test survey administration.

This contribution has been peer-reviewed.
https://doi.org/10.5194/isprs-archives-XLIV-M-2-2020-69-2020

4. FUTURE RECOMENDATIONS

4.1 Incorporate Relevant Data Sources

The Geospatial Data Act of 2018 will increase public and private access to geospatial data (The Federal Geographic Data Committee, 2020). Geospatial data is continuously being made available for public and private use. The inclusion of this new data at local and regional levels has the opportunity to produce better local-level mapping applications. For example, new geospatial data will be included with GeoPlatform.gov, the National Aeronautics and Space Administration, and the US Department of Defense. Additionally, with the current issue of COVID-19 throughout the US and globally, more options are opening up for students to collect and apply geospatial data

4.2 Integrate Advanced-Level Activities

As geospatial scientists, we know our field of study best. Therefore, we are equipped to expose students to advanced-level geospatial technologies that may not be offered through traditional K-12 systems. For example, once students are at ease using drone software, opportunities to utilize advanced-level drones, such as drones that follow the terrain like a missile or follow the user, may be explored with students and applied to exploratory research applications (e.g., emergency management, agriculture, food insecurity).

4.3 Discipline Specific Educational Research

Discipline specific educational research and K-12 student engagement are a natural fit. Students may provide collaborative insight, unexpected feedback and suggest technological applications that have never been considered or presented through formal research avenues. Understanding these variabilities are critical for geospatial science and education that extends beyond educational classrooms. Ultimately, every geoscientist's journey begins with exposure and opportunity because it is impossible for a student to identify with and express in interest in a field of study in which they have never shown as existing.

4.4 Inclusion of Supplemental Activities

The inclusion of supplemental activities are helpful when designed to expose students to new geospatial technologies. Supplemental activities related to the current research may be classified as research-based or relaxation-based. Research-based topics include machine learning, agent-based modeling, simulations, smart and precision agriculture, flood management, and renewable energy siting. Relaxation-based activities include geocaching, fishing, scavenger hunts, yoga, line-dancing, and meditation.

ACKNOWLEDGEMENTS

This material is based upon work supported by Title III and the National Science Foundation under Grants #1800768 and #1832110. Any opinions, findings, and conclusions or recommendations expressed in this material are those of the author(s) and do not necessarily reflect the views of the National Science Foundation.

COMPETING INTERESTS STATEMENT

The authors have no competing interests to declare.

REFERENCES

American Association for the Advancement of Science, 2020. *What are geospatial technologies?.* [Online]
Available at: https://www.aaas.org/programs/scientific-responsibility-human-rights-law/overview-geospatial-project
[Accessed 1 June 2020].

American Geoscience Institutue , 2010. *K-12 Teachers and Geoscience Degrees.* [Online] Available at: https://www.americangeosciences.org/geoscience-currents/k-12-teachers-and-geoscience-degrees
[Accessed 1 July 2020].

Archer, R. S., Davis, F., Ebanks, S. C. & Schulterbrandt Gragg III, R. D., 2018. HBCUs Broadening Participation in Geosciences (A Journey Through InTeGrate). In: *Interdisciplinary Teaching About Earth and the Environment for a Sustainable Future. AESS Interdisciplinary Environmental Studies and Sciences Series.* New York City: Springer, Cham, pp. 361-378.

Ayotte-Beaudet, J., Potvin, P., Lapierre, H. & Glackin, M., 2017. Teaching and Learning Science Outdoors in Schools' Immediate Surroundings at K-12 Levels: A Meta-Synthesis. *EURASIA Journal of Mathematics Science and Technology Education,* pp. 5343-5363.

Cagle, N. L., Caldwell, L. & Garcia, R., 2018. K-12 Diversity Pathway Programs in the E-STEM Fields: A Review of Existing Programs and Summary of Perceived and Unmet Needs. *Journal of STEM Education,* pp. 12-18.

Federal Geographic Data Committee, 2016. *Emerging Technologies and the Geospatial Landscape: A Report by the National Geospatial Advisory Committee. Available online: https://www. fgdc. gov/ngac/meetings/dec-2016/ngac-paper-emerging-technologies-and-the. pdf (accessed on 16 June 2020),* s.l.: s.n.

Fenichel, M. & Schweingruber, H. A., 2010. *Surrounded by Science: Learning Science in Informal Environments. Board of Science Education, Center for Education, Division of Behavioral and Social Sciences and Education,* Washington, DC: The National Academies Press.

Hammond, T. et al., 2019. Shoulder to Shoulder: Teacher Professional Development and Curriculum Design and Development for Geospatial Technology Integration With Science and Social Studies Teachers. *Contemporary Issues in Technology and Teacher Education,* 19(2), pp. 279-301.

Hawthorne, T. L. & Jarrett, O. S., 2018. Developing the Next Generation of CommunityBased Scholars. *The Professional Geographer,* pp. 291-297.

Huntoon, J., Peach, C. & Hopkins, J., 2005. *Geoscience Education and Diversity: Vision for the Future and Strategies for Success Report of the Second Geoscience Education Working Group,* Washington, DC: Nationla Science Foundation Report.

Keane, C., 2018. *American Geosciences Institute.* [Online] Available at: https://www.americangeosciences.org/sites/default/files/currents/Currents-130_0.pdf [Accessed 2 July 2020].

Montano, P., 2016. *A Start To Developing a Logic Model.* [Online] Available at: https://www.informalscience.org/news-views/start-developing-logic-model

National Academies of Sciences, Engineering, and Medicine, 2019. *Science and Engineering for Grades 6-12: Investigation and Design at the Center,* Washington, DC.: The National Academies Press.

National Research Council, 2009. *Learning Science in Informal Environments: People, Places, and Pursuits.* Washington: DC: THe National Academies Press.

National Research Council, 2013. *Future U.S. Workforce for Geospatial Intelligence,* Washington, DC: The National Academies Press.

Orion, N., 2019. The future challenge of Earth science education research. *Disciplinary and Interdisciplinary Science Education Research,* pp. 1-8.

Osborne, Z. M., van de Gevel, S. L., Eck, M. A. & Sugg, M., 2020. An Assessment of Geospatial Technology Integration in K-12 Education. *Journal of Geography,* pp. 12-21.

Petcovic, H. L. et al., 2018. Research on Elementary, Middle, and Secondary Earth and Space Sciences Teacher Education. In: *A Community Framework for Geoscience Education Research.* Northfield: National Association of Geoscience Teachers, pp. 35-48.

Roberts, T. et al., 2018. Students' perceptions of STEM learning after participating in a summer informal learning experience. *International Jouranl of STEM Education,* 5(35), pp. 1-14.

Seymour, E. & Hunter, A.-B., 2019. *Talking about Leaving Revisited: Persistence, Relocation, and Loss in Undergraduate STEM Education.* Chan,, Switzerland: Springer.

The Federal Geographic Data Committee, 2020. *Geospatial Data Act of 2018.* [Online] Available at: https://www.fgdc.gov/gda[Accessed 15 June 2020].

U.S. Department of Labor, Employment and Training Administration, 2020. *Geospatial Technology Competency Model.* [Online]Available at: https://www.careeronestop.org/CompetencyModel/competency-models/geospatial-technology.aspx [Accessed 7 June 2020].

SPATIAL DIFFERENCES IN FRESH VEGETABLE SPENDING: A CASE STUDY IN GUILFORD COUNTY, NORTH CAROLINA

Haoran Miao[1]*, Leila Hashemi-Beni[1], Timothy Mulrooney[2], Lyubov A. Kurkalova[1], Chyi-Lyi Liang[1], Manoj Jha[1], Gregory Monty[1]

[1] North Carolina A&T State University
[2]North Carolina Central University
*hmiao@ncat.edu

KEY WORDS: Fresh Vegetable Consumption, Food Desert, Agent-based Model, Local Food System, Platform

ABSTRACT:

This paper investigates the spatial differences in fresh vegetable spending in Guilford County, North Carolina. We create a geo-coded spatial-temporal database for both human factors and natural factors to understand why food deserts have become a serious issue in a county with many farming activities. We find that residents living in food deserts do not buy enough fresh vegetables compared with their counterparts, even when they are shopping at full-service grocery stores. Social-economic factors are most sensitive and are important determinants of fresh food demand. Using an agent-based toy model, we find that fresh vegetable demand in each census tract in Guilford County varies to a large extent. The results suggest that the formation of food deserts may root from the demand side.

1. INTRODUCTION

Vegetables are important in a healthy diet. However, only ten percent adult Americans meet the vegetable consumption standard recommended by Dietary Guidelines for Americans and vegetable is the most under-consumed nutritional food in the States (2018 State Indicator Report on Fruits and Vegetables). There is growing literature in enhancing food security, especially providing sufficient fresh vegetables to communities through local food systems (e.g., Torjusen, Lieblein &Vittersø 2008; Wilkins, Farrell & Rangarajan 2015). Local food systems can improve food access (Kantor 2001), decrease carbon footprint (Kaplin 2011), reduce consumers' energy intake (Rose et al. 2008), ease the food desert problem (McKinney and Kato, 2017), one type of geographic disparities in the food supply. We propose to mitigate the food desert problem by using local food systems from a coupled human and natural system perspective. We are particularly focusing on improving the availability and accessibility of fresh vegetables in the food desert area by making good use of local food systems. This research takes the first step to investigate spatial differences in fresh vegetable demand. We use Guilford County, North Carolina (N.C.) as a case study.

A food desert is a metaphor for neighborhood health food deprivation. It has many versions of definition. In general, it refers to areas with low access to affordable fresh vegetables and fruits, usually in the units of census tracts. The United States Department of Agriculture (USDA) defines one census tract as a food desert if it meets two thresholds: 1) "a poverty rate of 20 percent or greater, or a median family income at or below 80 percent of the statewide or metropolitan area median family income"; 2) "at least 500 persons and/or at least 33 percent of the population lives more than 1 mile from a supermarket or large grocery store (10 miles, in the case of rural census tracts)". In other words, one measurement of food deserts by USDA is whether most residents in one census tract have nearby access to a full-service grocery store, which serves a variety of fresh vegetables and fruits, such as a supermarket or a wholesale market. Based on recent data (USDA ERS 2010, 2015; U.S. Census Bureau 2017), we find the food deserts in N.C. are expanding, and the number of full-service grocery stores is declining. Some solutions were taken by policymakers to ease the food desert problem. But very few cases had meaningful effects. For example, In Greensboro, N.C., Renaissance Co-op opened as a food desert rescuer, but they were out of business in two years due to a lack of enough demand.

Guilford County, N.C., has a population of more than half a million. It is divided into about 120 census tracts, and 21 of them are defined as food deserts by USDA. In Figure 1a, we find that those food deserts are surrounded by the undeveloped area where exists about 854 farms (USDA 2017, Census of Agriculture), and those farms are capable of producing a large variety of fresh vegetables.

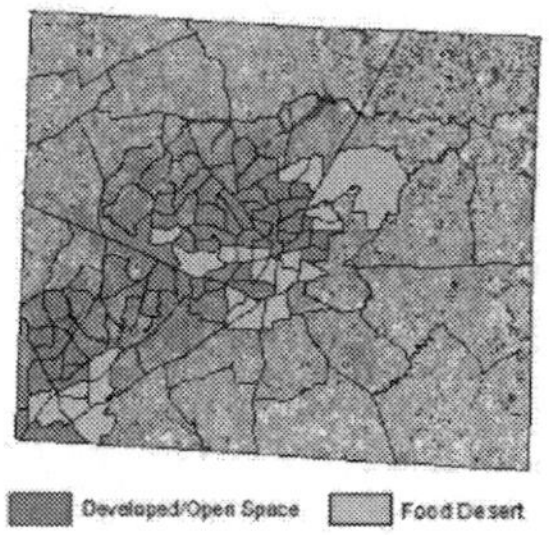

Figure 1a: Food deserts surrounded by undeveloped area/farms (represented by colors other than gray and pink) in Guilford County, North Carolina. *Data source: USDA food desert mapping and USDA Cropland Data Layer*

If we further overlay individual household addresses (RTI international's U.S. synthetic household population) on the map

*Corresponding author

https://doi.org/10.5194/isprs-archives-XLIV-M-2-2020-73-2020

(see Figure 1b), we may find that a significant portion of households living in food desert areas. USDA estimates that about one-fifth of the county's population is underserved by full-service grocery stores in Guilford County.

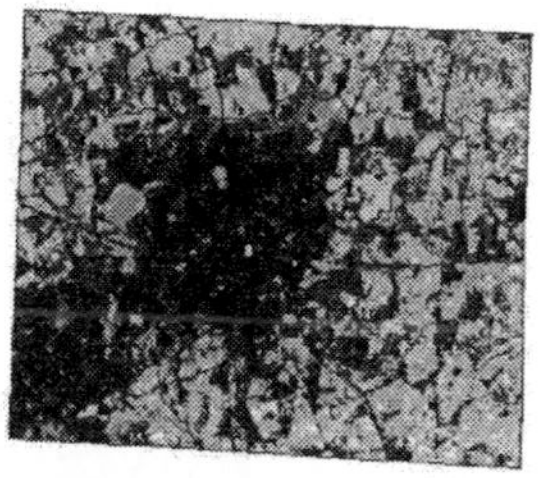

Figure 1b: Food deserts surrounded by undeveloped area/farms (represented by colors other than gray and pink) and households scattered in Guilford County. *Data source: USDA food desert mapping, USDA Cropland Data Layer, RTI international's U.S. synthetic household population*

One important difference between one full-service grocery store and one corner store is the different supply of fresh vegetables and fruits. As we know, supply and demand always come side by side, and most literature argues that the food desert area (e.g., Allcott et al. 2019) is underserved mostly because of the low demand for healthy food. However, we find that the current literature and data mostly focuses on finding the association between food environment/food deserts and dietary behavior/healthy food at the individual level. For example, the U.S. Bureau of Labor Statistics (BLS) conducts consumer expenditure surveys every year, which gives average spending on fresh vegetables pre household. From the dataset, we can draw a map that displays the correlations between food deserts and the low consumption level of fresh vegetables. Figure 2 demonstrates that households living in most food desert areas consume fresh vegetables less than $210 per year on average.

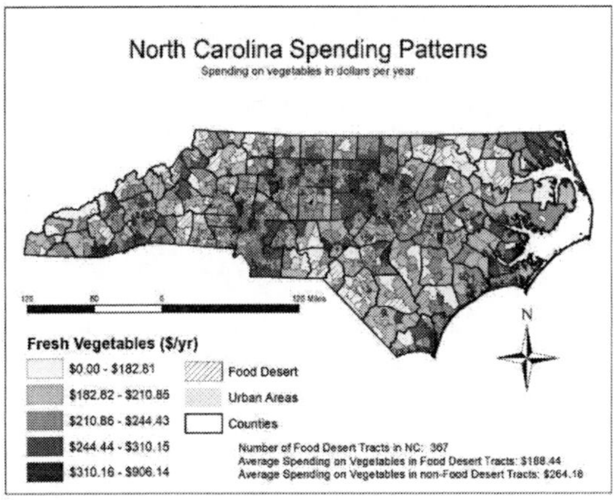

Figure 2: Spatial differences in fresh vegetable consumption in North Carolina. *Data Sources: ESRI's 2019 U.S. Consumer Spending Database (originally from BLS) and USDA Food Desert Mapping*

However, there is a lack of literature that appropriately calculate the spatial differences in vegetable spending between food-desert and non-food desert area at the census tract level. The gap in the literature leads to our first step research question: what the spatial differences in fresh vegetable spending between food-desert and non-food-desert census tracts in Guilford County are. We propose to use a combination of a private dataset and several public datasets to solve this problem. The technique to calculate the aggregated fresh vegetable consumption is agent-based modeling, which is a from-bottom-to-up approach. In general, we start from household level and use agents in the software NetLogo to represent about 220k households of Guilford County. Their fresh vegetable purchasing behavior is affected by the food environment and their household features, which are programmed as one set of fresh vegetable purchasing rules in the NetLogo. We use the software to simulate each household behavior for one year, and then we aggregate fresh vegetable demand for each census tract.

2. DATA

The dataset that we first look at in our study area is the Cropland Data Layer (CDL) from the USDA National Agricultural Statistics Service (NASS). Using images remotely sensed by satellites and national agricultural statistics, USDA NASS creates the CDL dataset annually to illustrate how the U.S. continents are covered by specific crops, includes different types of vegetables. CDL data contain one layer of raster data that show how specific crops are distributed over space and over time. To fulfill the purpose of the paper, we mainly utilize the CDL dataset in the year 2019 to find out how the current spatial relationships among vegetable growing parcels, household addresses, and food desert areas. For the next steps, we would apply CDL data in about ten years to estimate how land use, especially vegetable growing area, changes over space and time, and whether they are correlated with farm-level characteristics and uncertain scenarios, such as a flood.

The second dataset that we use for our study area is the RTI international's U.S. synthetic household population dataset (Wheaton et al. 2009). Because we use a bottom-up strategy to estimate how fresh vegetable demand is spatially different using an agent-based model and socioeconomic differences matter in the correlation between food deserts and dietary behavior (Mackenbach et al. 2019), we need data containing the spatial distribution of households and the characteristics of each household in our study area. Different from aggregated data at census tracts or zip code levels, The RTI international's U.S. synthetic household population dataset represents an accurate and complete set of household addresses and household (member) characteristics, such as household income, member ages, race, household size, etc. Therefore, in the agent-based model, one household, represented by one agent, can follow specific rules and make fresh vegetable purchasing decisions based on their household-specific data (e.g., whether one household locates in food desert area). Also, we can intuitively understand the spatial relationships between the food desert area/fresh vegetable production area and household locations.

The third dataset, based on which we extract behavior patterns, is the Nielsen Homescan dataset. Nilsen Homescan data are one type of national-level dataset provided by Nielsen company. The company has a balanced sample all over the States, and each panelist in the sample reports all purchased items, including all kinds of fresh vegetables. We can observe the types of vegetables, unit price, and total spending of one vegetable item from the dataset. Also, we can observe the types of stores where those vegetables are bought from, such as a wholesale club, a supermarket, or a convenience store.

Furthermore, each panelist reports the location of his/her household, and household characteristics, such as household

This contribution has been peer-reviewed.
https://doi.org/10.5194/isprs-archives-XLIV-M-2-2020-73-2020

income, household member ages, employment status, race, etc. We use the subset of N.C. so that we can better mimic the fresh vegetable purchasing behavior in our study area. The dataset records panelists' each grocery shopping trip, and each item they bought during one trip. We also combine the Nielsen Homescan data and USDA food desert data to estimate the food desert status.

One important feature of Nielsen Homescan data is that the dataset only discloses panelists' geographic location at the zip code level to protect the privacy of the panelists. However, the USDA census tract definition is based on the census tract level. We use the fourth dataset to estimate the food desert status of one zip code: the HUD-USPS ZIP Crosswalk. This dataset is organized by the U.S. Department of Housing and Urban Development's (HUD's) Office of Policy Development and Research (PD&R). It estimates how many residents living in a census tract live in a zip code. We define one zip code overlapping with any food-desert census tracts as a food-desert zip code.

3. METHOD

We aim to build an integrated agent-based model to achieve our goal for our whole project. The overarching goal of the agent-based model is to build a food system that includes vegetable production and consumption, and related environmental impact. Thus, the agent-based model includes three types of agents, households, farms, and environmental agents. We identify their rules of the agents from our datasets and previous literature.

Following Abel and Faust (2020), we assume households make one to four decisions each tick or day based on their food environment and household characteristics: whether they are going to do grocery shopping, which store they are going to do grocery shopping, whether they are going to buy fresh vegetables, and how much they are going to spend on fresh vegetables. By analyzing Nielsen Homesman data, we can find the parameters for fresh vegetable purchasing behavior.

As farm agents, they decide whether they change the use of land at the beginning of one season, either turn vegetable planting into other uses, or vice versa. Based on the previous years' CDL data, we model the change process as functions of changes in previous years' consumption and changes in vegetable prices, and other variables that have significant effects on land-use change. The environmental agents would take variables from farm agents' actions, such as land-use change, to simulate environment changes, such as water quality, using Agricultural Policy/Environmental eXtender (APEX model). The three parts, production, consumption, and environment, would be linked and integrated by variables such as vegetable prices, vegetable yield, and weather conditions.

However, to answer the research question of this paper, we only initialize and parametrize the production model part in the agent-based model to calculate the spatial differences of fresh vegetable consumption in Guilford county. We will initialize and parametrize other parts of the agent-based model in our future work.

4. RESULTS

Table 1 presents some descriptive statistics of our sample data from Nielsen Homescan data. We also compare those variables based on their food desert status. In the dataset, vehicle ownership and age are not statistically different between non-food-desert and food-desert households. But, other variables, such as income, education year, household size, are sharply different. For example, the average household income of food-desert samples is $57,000 (2010 U.S. dollars), which is about $8,000 lower than that of their counterparts. From the descriptive statistics, we also find that a large portion of minorities may live in a food desert area.

Variables	Non-food deserts			Food deserts			Balance test (χ^2)
	Median	Mean	SD.	Median	Mean	SD.	*p*-value
Vehicle	1	0.70	0.45	1	0.71	0.45	0.45
Income	55.52	64.24	35.89	49.50	56.98	33.66	0.00***
Education years	15	14.70	1.86	14	14.53	1.89	0.00***
Household size	2	2.50	1.24	2	2.42	1.25	0.00***
Age	55	53.98	12.18	55	54.32	12.07	0.108
Hours	20	22.81	15.40	20	21.95	15.97	0.00***
White	1	0.83	0.37	1	0.71	0.45	0.00***
Black	0	0.10	0.30	0	0.24	0.43	0.00***
Asian	0	0.02	0.14	0	0.01	0.12	0.00***
Hispanic	0	0.04	0.19	0	0.03	0.16	0.03**
Married	1	0.73	0.44	1	0.66	0.47	0.02**
Child	0	0.25	0.44	0	0.23	0.42	0.00***
Employed	0.50	0.62	0.41	0.50	0.59	0.42	0.00***

Table 1. Descriptive statistics of household characteristics by food desert status

*Notes: Sample size for non-food-desert residents is 2,748, sample size for residents living in or close to a food desert is 5,891. Total sample includes 8,640 household-by-year observations for 2012-2017 in North Carolina. Vehicle = 1 if owning a vehicle, = 0 otherwise; Income: household income ($1000s); Education years: Average of education years of household head(s); Household size: member numbers of a household; Age: Average of ages of household head(s); Hours: Average of weekly working hours of household head(s); White = 1 if white households, = 0 otherwise; Black = 1 if black households, = 0 otherwise; Asian = 1 if Asian households, = 0 otherwise; Hispanic = 1 if Hispanic households, = 0 otherwise; Married = 1 if married households, = 0 otherwise; Child = 1 if any child under 18 years old in the households, = 0 otherwise; Employed = 0 if no head employed; = 0.5 if one of the heads employed in a two heads household; = 1 if head(s) employed. *** $p<0.01$, ** $p<0.05$, * $p<0.1$.*

After analyzing the data systematically, we can conclude a few consumption patterns. Firstly, the numbers of total trips in a year and the numbers of trips to a full-service grocery store are not different from each other between food desert households and their counterpart. For example, during a year, both food-desert and non-food-desert households visit full-service grocery stores 92 times on average, and the differences are not statistically significant (T-Test, $p > 0.1$.).

If we take other confounders together and try to understand how they affect the probabilities of households going to a full-service grocery store, we can see the ownership of a vehicle is the most powerful indicator (Table 2). It decreases the probability of choosing a full-service grocery store by about 15%. Intuitively, one household will have many choices if they own vehicles. In other words, although families without cars may decrease their grocery shopping trips, they would go for shopping at full-service grocery stores at greater probabilities than those having cars. The

This contribution has been peer-reviewed.
https://doi.org/10.5194/isprs-archives-XLIV-M-2-2020-73-2020 |

food environment (food desert status) matters, but the magnitude seems small. Households with children would have a 4% higher probability to go for shopping at full-service grocery stores, ceteris paribus. The ethnic groups also play an important role in choosing grocery stores.

F.D.	V	I	E	H	A	B	AS	C
-1.54%***	-15.50%***	0.06%***	0.99%***	0.46%***	0.06%***	-5.50%***	4.90%***	3.63%***
-0.00093	-0.0011	-0.000015	-0.00024	0.00055	-0.000004	-0.002	-0.0045	-0.0015

Table 2. Determinants of choosing full-service grocery stores
*Notes: Logit regression results for the prob. of choosing a full-service grocery store, N = 1,107,754. First row: Marginal probability. Second row: Delta-method standard errors. *** p<0.01, ** p<0.05, * p<0.1. Abbreviations: FD = Food Desert (= 1 if living in a food desert, = 0 if not living in a food desert); V = Vehicle; I = Income; E = Education years; H = household size; A = Age; B = Black; AS = Asian; C = Child.*

Secondly, we find that most vegetables are bought in full-service grocery stores. Households spend about 98% of their fresh vegetable budget at full-service grocery stores, no matter where they live. After arriving at a full-service grocery store, patterns to buy fresh vegetables are different across different household characteristics. The logistic regression results (Table 3) show that owning a vehicle does affect the probabilities of buying vegetables. But living in a food desert and being an African American would decrease the probability of buying fresh vegetables statistically significantly. The probabilities of buying fresh vegetables at full-service grocery stores increase as the household income and education years increase.

F.D.	V	I	E	H	A	B	AS	C
-2.17%***	-0.11%	0.034%***	0.78%***	0.55%***	0.094%***	-4.12%***	3.24%***	0.11%
-0.00112	-0.0012	-0.000018	-0.00031	0.00069	-0.000059	-0.0031	-0.0047	-0.0019

Table 3. Determinants of buying vegetables at a full-service grocery store
*Notes: Logit regression results for the probability of choosing a full-service grocery store, N = 1,107,754. First row: Marginal probability. Second row: Delta-method standard errors. *** p<0.01, ** p<0.05, * p<0.1. Abbreviations: FD = Food Desert (= 1 if living in a food desert, = 0 if not living in a food desert); V = Vehicle; I = Income; E = Education years; H = household size; A = Age; B = Black; AS = Asian; C = Child.*

Thirdly, we find that the fresh vegetable spending during each trip is different if we compare the two groups. Non-food-desert households buy 4.35-dollar fresh vegetables on average in each trip, while the number for their counterparts is 3.95 dollars. If we add impacts of other confounders by running a weighted least square regression (Table 4), we can find that the impact of food deserts is still statistically significant. The fresh vegetable spending increases as the increases in the household income and education year. Minorities groups do not have a big difference in consumption if they arrive at a full-service grocery store and decide to buy vegetables.

F.D.	V	I	E	H	A	B	AS	C
-0.32***	0.0045	0.0058***	0.050*	0.19***	-0.018***	-0.27	0.07	-0.06
-0.12	-0.1	-0.0018	-0.029	(0.058)	-0.0051	-0.26	-0.26	-0.04

Table 4. Determinants of the spending on fresh vegetables
*Notes: Weighted least square regression results, N = 264,582. First row: Marginal fresh vegetable spending. Second row: robust standard errors clustered at household level. *** p<0.01, ** p<0.05, * p<0.1. Abbreviations: FD = Food Desert (= 1 if living in a food desert, = 0 if not living in a food desert); V = Vehicle; I = Income; E = Education years; H = household size; A = Age; B = Black; AS = Asian; C = Child.*

After we implement those above patterns and the trigger of going for grocery shopping (a probability function similar in Abel and Faust, 2020) in the agent-based model and simulate agents' behavior with other environmental data, such as household food environment, and household characteristics, we can find the spatial differences in yearly fresh vegetable consumption in Guilford County (Table 5). The simulation results show that, during a year, a non-food-desert census tract consumes about 184k dollars of fresh vegetables on average, while a food desert census tract consumes about 124k dollars. Non-food-desert census tracts consumers about 50% more than their counterparts.

Non-food-desert census tracts			Food desert census tracts			T-test
Median ($)	Mean ($)	S.D.	Median ($)	Mean ($)	S.D.	p-value
177K	184K	84K	106K	124K	75K	0.00***

Table 5. Spatial differences in yearly fresh vegetable consumption in Guilford County

5. DISCUSSION AND FUTURE STEPS

In this paper, we propose a way to build an integrated model of a local food system using the agent-based model technique. We take the first step to estimate the spatial differences in fresh vegetable spending in Guilford County, NC, using a part of the model. Combing a few private and public datasets, we can estimate aggregated fresh vegetable spending at the census tract level. We find that census tracts defined as food deserts by USDA have a much lower fresh vegetable demand, which indicates Food deserts may be equilibrium responses to consumers' demand.

With the mentioned datasets and proposed method, we built up a platform to estimate the spatial fresh vegetable demand. We can easily extend the method to other food, such as fresh fruits, canned vegetables, or prepared food. We believe the proposed method would help future research about food security. One limitation of the current platform is that we do not validate our results with data from other resources. We will plan to do that for our next step.

In the future, we will develop one more integrated model, which includes consumers, farmers, and environmental agents in the model. More data, such as remote sensing data, GIS info, land use data, will also be included in the model as environment data. Ultimately, we want to find out to what extent how the local food system can help alleviate the food desert problem, and any environmental consequences would happen.

ACKNOWLEDGEMENT

This material is based upon work supported by the National Science Foundation under Grant No. 1824949. Any opinions, findings, and conclusions or recommendations expressed in this material are those of the author(s) and do not necessarily reflect the views of the National Science Foundation. Researcher(s) own analyses calculated (or derived) based in part on data from The Nielsen Company (U.S.), LLC, and marketing databases provided through the Nielsen Datasets at the Kilts Center for Marketing Data Center at The University of Chicago Booth School of Business. The conclusions drawn from the Nielsen data are those of the researcher(s) and do not reflect the views of Nielsen. Nielsen is not responsible for, had no role in, and was not involved in analyzing and preparing the results reported herein.

REFERENCES

"2014-Economics-of-Local-Food-Systems.Pdf." Accessed July 11,2020. https://conservancy.umn.edu/bitstream/handle/11299/171637/2014-Economics-of-Local-Food-Systems.pdf?sequence=1.

"2018 State Indicator Report on Fruits and Vegetables," n.d., 18.
Allcott, Hunt, Rebecca Diamond, Jean-Pierre Dubé, Jessie Handbury, Ilya Rahkovsky, and Molly Schnell. "Food Deserts and the Causes of Nutritional Inequality*." The Quarterly Journal of Economics 134, no. 4 (November 1, 2019): 1793–1844. https://doi.org/10.1093/qje/qjz015.

Abel, Kelsey C., and Kasey M. Faust. "Modeling Complex Human Systems: An Adaptable Framework of Urban Food Deserts." Sustainable Cities and Society 52 (January 1, 2020): 101795. https://doi.org/10.1016/j.scs.2019.101795.
Dong, Diansheng, and Biing-Hwan Lin. "Fruit and Vegetable Consumption by Low-Income Americans: Would a Price Reduction Make a Difference?," n.d., 23.

Kantor, Linda Scott. "Community Food Security Programs Improve Food Access." Food Review/ National Food Review, January 1, 2001. https://doi.org/10.22004/ag.econ.266234.

Kaplin, Lauren. "Energy (In)Efficiency of the Local Food Movement: Food for Thought" 23 (n.d.): 24.

Mackenbach, Joreintje D., Kyra G. M. Nelissen, S. Coosje Dijkstra, Maartje P. Poelman, Joost G. Daams, Julianna B. Leijssen, and Mary Nicolaou. "A Systematic Review on Socioeconomic Differences in the Association between the Food Environment and Dietary Behaviors." Nutrients 11, no. 9 (September 2019): 2215. https://doi.org/10.3390/nu11092215.

Markowitz, Lisa. "Produce(Ing) Equity: Creating Fresh Markets in a Food Desert." In Hidden Hands in the Market: Ethnographies of Fair Trade, Ethical Consumption, and Corporate Social Responsibility, edited by Geert De Neve, Luetchford Peter, Jeffrey Pratt, and Donald C. Wood, 28:195–211. Research in Economic Anthropology. Emerald Group Publishing Limited, 2008. https://doi.org/10.1016/S0190-1281(08)28009-1.

McClintock, Nathan, Jenny Cooper, and Snehee Khandeshi. "Assessing the Potential Contribution of Vacant Land to Urban Vegetable Production and Consumption in Oakland, California." Landscape and Urban Planning 111 (March 1, 2013): 46–58. https://doi.org/10.1016/j.landurbplan.2012.12.009.

McKinney, Laura, Yuki Kato, 1 Department of Sociology, Tulane University, USA, and 2 Department of Sociology, Georgetown University, USA. "Community Context of Food Justice: Reflections on a Free Local Produce Program in a New Orleans Food Desert." AIMS Agriculture and Food 2, no. 2 (2017): 183–200. https://doi.org/10.3934/agrfood.2017.2.183.

Morrison, Kathryn T., Trisalyn A. Nelson, and Aleck S. Ostry. "Mapping Spatial Variation in Food Consumption." Applied Geography, Local Food Systems and the Applied Geography of Food, 31, no. 4 (October 1, 2011): 1262–67. https://doi.org/10.1016/j.apgeog.2010.11.020.

MS, Nick Rose, Elena Serrano PhD, Kathy Hosig PhD RD MPH, Carola Haas PhD, Dixie Reaves PhD, and Sharon M. Nickols-Richardson PhD R.D. "The 100-Mile Diet: A Community Approach to Promote Sustainable Food Systems Impacts Dietary Quality." Journal of Hunger & Environmental Nutrition 3, no. 2–3 (August 25, 2008): 270–85. https://doi.org/10.1080/19320240802244082.

Nolasco, Camille, Luciana Soler, Marcos Freitas, Myanna Lahsen, and Jean Ometto. "Scenarios of Vegetable Demand vs. Production in Brazil: The Links between Nutritional Security and Small Farming." Land 6, no. 3 (July 25, 2017): 49. https://doi.org/10.3390/land6030049.

Noy, Sue, Rebecca Patrick, Claire Henderson-Wilson, Sonia Nuttman, and Isabella Ryan. "New Frontiers in Community Initiatives to Increase Vegetable Consumption." Health Promotion Journal of Australia 30, no. S1 (2019): 52–61. https://doi.org/10.1002/hpja.207.

Osvald, Ana, and Lidija Zadnik Stirn. "A Vehicle Routing Algorithm for the Distribution of Fresh Vegetables and Similar Perishable Food." Journal of Food Engineering 85, no. 2 (March 1, 2008): 285–95. https://doi.org/10.1016/j.jfoodeng.2007.07.008.

Peters, Christian, Nelson L. Bills, Jennifer Wilkins, and R. David Smith. "Vegetable Consumption, Dietary Guidelines and Agricultural Production in New York State—Implications for Local Food Economies." AgEcon Search, 2002. https://doi.org/10.22004/ag.econ.122636.

Torjusen, Hanne, Geir Lieblein, and Gunnar Vittersø. "Learning, Communicating and Eating in Local Food-Systems: The Case of Organic Box Schemes in Denmark and Norway." Local Environment 13, no. 3 (April 15, 2008): 219–34. https://doi.org/10.1080/13549830701669252.

Wheaton, William D., James C. Cajka, Bernadette M. Chasteen, Diane K. Wagener, Philip C. Cooley, Laxminarayana Ganapathi, Douglas J. Roberts, and Justine L. Allpress. "Synthesized Population Databases: A US Geospatial Database for Agent-Based Models." Methods Report (RTI Press) 2009, no. 10 (May 1, 2009): 905. https://doi.org/10.3768/rtipress.2009.mr.0010.0905.

Wilkins, Jennifer L., Tracy J. Farrell, and Anusuya Rangarajan. "Linking Vegetable Preferences, Health and Local Food Systems through Community-Supported Agriculture." Public Health Nutrition 18, no. 13 (September 2015): 2392–2401. https://doi.org/10.1017/S1368980015000713.176.
doi:https://doi.org/10.1016/j.landusepol.2015.07.008

EXTRACTING BUILT-UP FEATURES IN COMPLEX BIOPHYSICAL ENVIRONMENTS BY USING A LANDSAT BANDS RATIO

Alfred Homère Ngandam Mfondoum*[1, 2], Paul Gérard Gbetkom[3] Ryan Cooper [4], Sofia Hakdaoui[5], Mamane Barkawi Mansour Badamassi[6]

[1] StatsN'Maps, Private Consulting Firm, Dallas, Texas, USA. stats.n.maps.expertise@gmail.com
[2] Department of Geography, University of Yaoundé I, Yaoundé, Cameroon. Alfred.homere@gmail.com
[3]Department of Geography, University of Aix-Marseille, Marseille, France. Paul-gerard.gbetkom@etu.univ-amu.fr
[4] Erik Jonsson School of Engineering and Computer Science, University of Texas in Dallas, Dallas, Texas 75080, USA. Rdc190001@utdallas.edu
[5] Earth Observation Department, Geo-Biodiversity and Natural Patrimony Laboratory, Geophysics, Natural Patrimony and Green Chemistry Research Center, Scientific Institute, Mohamed V University, Rabat, Morocco. s.hakdaoui@gmail.com
[6]Laboratory of Botanic, Mycology and Environment, Mohammed V University, Rabat, Morocco. barkawimansour@yahoo.fr

Keywords: Built-up, urban mixed pixels, Normalized Difference Built-up and Surroundings Unmixing Index, Landsat-8, Yaoundé, Cameroon,

ABSTRACT:

This paper addresses the remote sensing challenging field of urban mixed pixels on a medium spatial resolution satellite data. The tentatively named Normalized Difference Built-up and Surroundings Unmixing Index (NDBSUI) is proposed by using Landsat-8 Operational Land Imager (OLI) bands. It uses the Shortwave Infrared 2 (SWIR2) as the main wavelength, the SWIR1 with the red wavelengths, for the built-up extraction. A ratio is computed based on the normalization process and the application is made on six cities with different urban and environmental characteristics. The built-up of the experimental site of Yaoundé is extracted with an overall accuracy of *95.51%* and a kappa coefficient of *0.90*. The NDBSUI is validated over five other sites, chosen according to Cameroon's bioclimatic zoning. The results are satisfactory for the cities of Yokadouma and Kumba in the bimodal and monomodal rainfall zones, where overall accuracies are up to *98.9%* and *97.5%*, with kappa coefficients of *0.88* and *0.94* respectively, although these values are close to those of three other indices. However, in the cities of Foumban, Ngaoundéré and Garoua, representing the western highlands, the high Guinea savannah and the Sudano-sahelian zones where built-up is more confused with soil features, overall accuracies of *97.06%*, *95.29%* and *74.86%*, corresponding to *0.918*, *0.89* and *0.42* kappa coefficients were recorded. Difference of accuracy with EBBI, NDBI and UI are up to *31.66%*, confirming the NDBSUI efficiency to automate built-up extraction and unmixing from surrounding noises with less biases.

1. INTRODUCTION

Every urban area is heterogeneous, mixing diverse land cover types and materials, such as road, rooftop, vegetated area, bare soil, water, which are quite different from rural and natural environment (Schueler, 1994). This usually leads to mixed pixels, known as a picture element that has a digital number representing the average energy emitted or reflected from several different surfaces occurring within the area represented, by the pixel (Bangira et al., 2017). Urban remote sensing offers several solutions to address these issues, mainly using spectral indices-based algorithms and decomposition methods.

Some simple indices based on ratio or algebraic crossings of basic spectral bands enable to automate the mapping of human settlements. It is the case of the urban index, UI (Kawamura et al., 1996), the normalized difference built-up index, NDBI (Zha et al., 2003) and the new built-up index, NBI (Chen et al., 2010). Some other focus on built-up and bare soil differentiation, as the Enhanced Built-Up and Bareness Index, EBBI (As-syakur, 2012), and the normalized difference tillage index, NDTI (Ettehadi et al., 2019).

Further, the composed indices based on thematic index-derived bands and existing algorithms modifications are interested in noise reduction. Amongst them are the index-based built-up index, IBI (Xu, 2008) and the dry built-up index, DBI (Rasul et al., 2018). To address this issue in terms of imperviousness, the normalized difference impervious surface index, NDISI (Hanqiu, 2010) and its derivatives were proposed.

All the above indices are helpful to highlight built-up areas. However, uncertainties and biases remain in separating bare land from built-up. The main reason is that spectral signatures of ISAs are like those of soils, sands and rocks. In developing countries, the extent of built-up in makeshifts and raw material, as well as unpaved (dusty/muddy) roads (Fig.1) inside and near cities, added to the use of freely available satellite images of coarse or moderate spatial resolution (MODIS, Landsat) in unfunded researches emphasize these issues. Consequently, there is a high spatial and spectral diversity of surface materials that are more challenging for remote sensing analysis (Herold et al., 2003).

* Corresponding author (stats.n.maps.expertise@gmail.com)

This contribution has been peer-reviewed.
https://doi.org/10.5194/isprs-archives-XLIV-M-2-2020-79-2020

Figure 1. A mixed built-up environment. *A=Bright aluminum roof; B= Rusted aluminum roof; C=Unpaved (Dusty/Muddy) road/Soils/Stones; D=Bright and rusted aluminum roofs mixed with bananas plants; E=Pavements mixed with stones and bare soils; F=Natural vegetation.*

From the above, the main goal of this research is to unmix built-up and surrounding land cover features, i.e. bare soils, vegetation and water. The two specific goals are the following: **i)** *extract the built-up features from a spectral bands ratio in mixed-housing and built-up material context;* **ii)** *dissociate interweaved spectral signals of built-up and natural surrounding features to map the urban area with the highest accuracy.*

2. METHODOLOGY

2.1. Research location

The experiments were conducted in Cameroon (Central Africa), located between latitude 2°00'0" - 13°00'0"N and longitude 8°00'0" - 16°00'0"E (Fig.2*a*.). The experimental site is Yaoundé, the political capital. This city belongs to the humid forest zone with bimodal rainfall, it is located between latitude 3°52'00.0"N and longitude 11°31'00.0"E (Fig. 2*b*.), its population is 2.5 million inhabitants and its area covers approximately 31750 hectares. Yaoundé is subject to urban sprawl due to an outdated master plan and a lack of further planning as other cities in Cameroon. Three areas of interest (AOIs) extracted in Yaoundé enable a good view of the sprawl as well as mixture. (Fig.2*b*)

For the purpose of cross validation, five other cities are inserted in the study. The cities of Yokadouma, Kumba, Foumban, Ngaoundéré and Garoua are selected according to Cameroon's bioclimatic zoning (Fig. 2.*c* to 2.*g*). Indeed, the specific conditions of high or low forest cover and consequently the related soil exposure, as well as altitude, should help to validate the suitability of the proposed index application.

2.2. Datasets acquisition and pre-processing

The data used are satellite images from Landsat 8 Operational Land Imager-Thermal infrared sensors (Tab.1). The scenes used were downloaded for the dry season from the United States Geological Survey website and displayed under false colour composite NIR-Red-Green (see fig.2*b*-*g*) (Tabl.1). For the purpose of pre-processing, bands blue, green, red, near infrared, and both shortwave infrareds were stacked. Applying the Cosine Solar TAUZ (COST) radiometric calibration model of Chavez (1996), the DNs were converted from at-sensor radiance to top-of-atmosphere (TOA) reflectance. Atmospheric corrections and haze reduction have helped to remove other noises and then approximate values of land surface reflectance. The last step concerned the topographic correction to lower altitude artifacts.

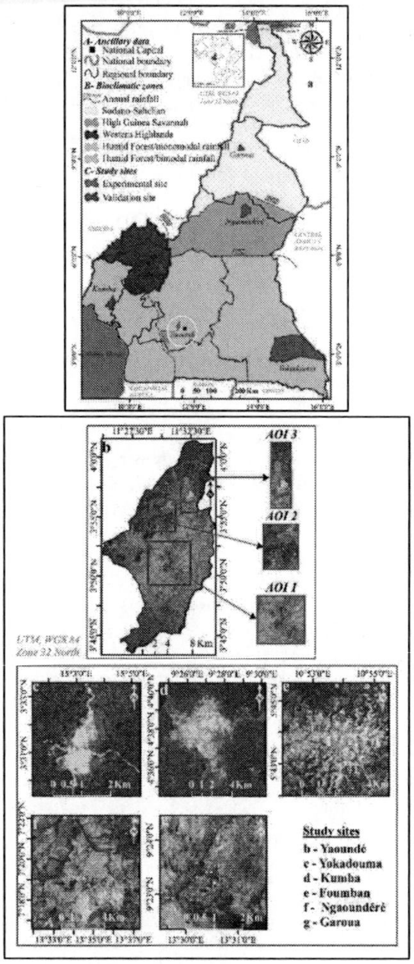

Figure 2. The study area

Site	*Scene ID*	*Center Time*
Yaoundé	LC81850572018004LGN00	09 : 26
Yokadouma	LC81820582019002LGN00	09 : 14
Kumba	LC81870572019005LGN00	09 : 38
Foumban	LC81860562019030LGN00	09 : 32
Ngaoundéré	LC81840552019064LGN00	09 : 19
Garoua	LC81850542019087LGN00	09 : 25

Table 1. Scenes information

2.3. Processing

2.3.1. The spectral separability efficiency of SWIR2

According to the experiences of Herold et al. (2003) and the USGS spectral library (Kokaly et al., 2017), there are several visible and infrared windows of the electromagnetic spectrum where manmade settlements and artificial surfaces reflect the best. The most suitable for the built-up separability from surroundings are in the

SWIR region located at *1710*, *1740*, *1960*, *2000*, *2040*, *2200*, *2340* and *2460* nanometers. To assess this statement on the study sites, spectral curves of main land cover objects were extracted, i.e. built-up, vegetation, wet soils/Swamps, dry soils/rocks and water bodies. (Fig.3)

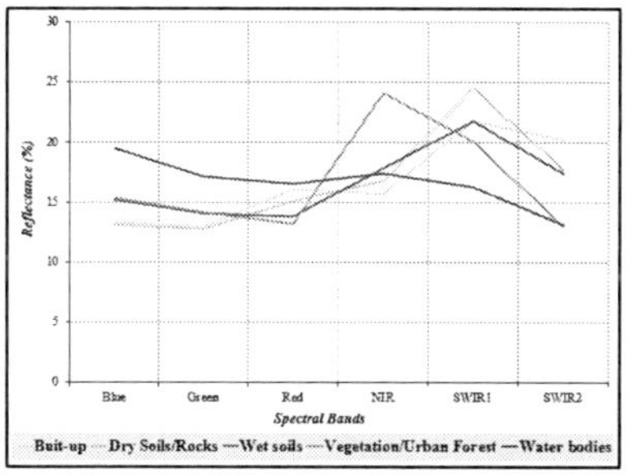

Figure 3. Spectral curves in Yaoundé

It can be noticed that the high unmixed reflectance of built-up takes place in the SWIR2 band (2107-2294 nm), its highest reflectance is in the SWIR1 band (1566- 1651 nm), but covered by the soil features ; while the red band (636-673 nm) shows the best built-up reflectance in the visible wavelengths despite the water bodies noise. A thorough assessment was performed through pixels sampling on the three areas of interest. Regardless of the dominating land cover class, on 10,000 pixels and from the stretching approach, the lower frequencies (counts) of the red band and the medium for the SWIR1 correspond to higher for SWIR2. (Fig.4 left column). While on 2,500 pixels, the values of SWIR2 digital numbers, DNs, are discriminated from the two other bands that are higher or lower (Fig.4; right column)

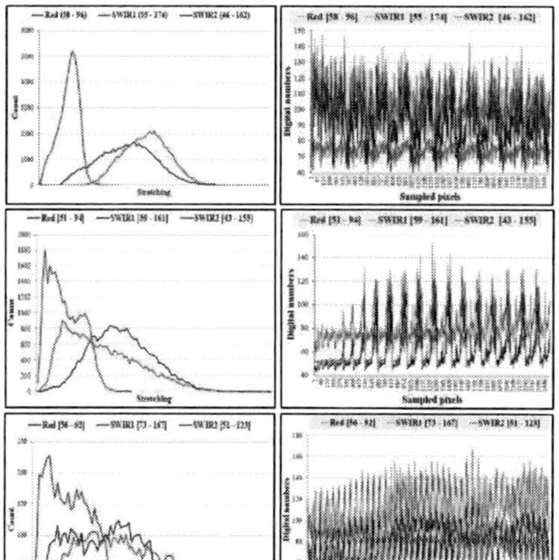

Figure 4. Status of red, SWIR1 and SWIR2 discrimination.

2.3.2. Built-up features extraction from the red and SWIR1 bands

The built-up information in the red and SWIR1 is important, but highly mixed with other features. The intent is to create a new pixel representing the built-up in these two bands and add it to the SWIR2 built-up information.

The approach through the geometric mean was chosen, because in image processing, the geometric mean filter is meant to smooth and reduce noise (Gonzalez, 2002). The output image G(*x*,*y*) is given by:

$$G(x,y) = \left[\prod_{i,j\in S} S(i,j)\right]^{\frac{1}{mn}} \quad (1)$$

Where S(*x*,*y*) is the original image, and the filter mask is *m* by *n* pixels. Each pixel of the output image at point (*x*,*y*) is given by the product of the pixels within the geometric mean mask raised to the power of $\frac{1}{mn}$ (Gonzalez, 2002).

The graphical representation and mathematical expression of the geometrical mean that eases the red and SWIR1 implementation is the following: (Fig.5)

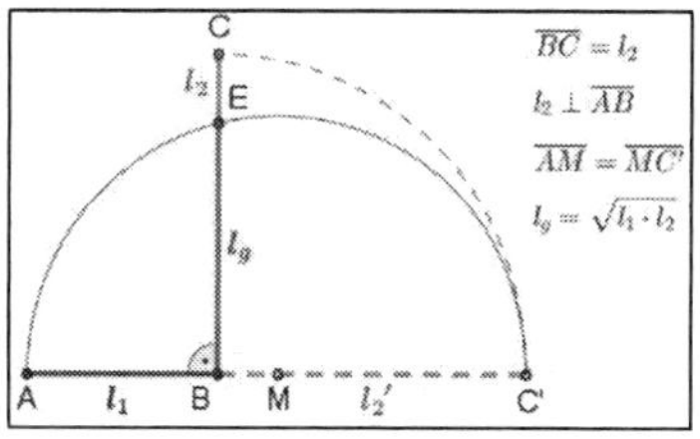

Figure 5. The geometric mean illustration

In addition, while proposing the EBBI algorithm, As-syakur (2012) recognized that a root function enables to cluster the numbers that contrast identical objects based on the different levels of reflectance values. Therefore, the equation for the new pixel was posed as follows:

$$G(\boldsymbol{Red}, \boldsymbol{SWIR1}) = \sqrt{\boldsymbol{Red} * \boldsymbol{SWIR1}} \quad (2)$$

Where Red is l_1, according to its lower built-up information than SWIR1 which considered l_2.

The sampling of built-up pixels on the stacked images confirms the value of the pixel created to be newly informative. (Fig.6)

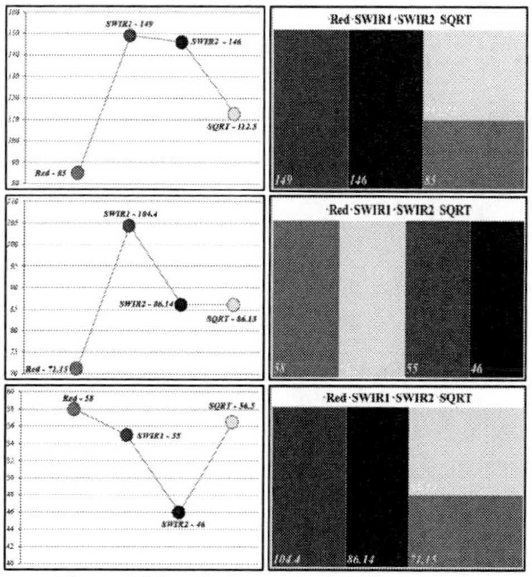

Figure 6. Three built-up pixels (left) and their informative values (right).

When the SWIR2 is regressed by the red, SWIR1 and the new band, the determination coefficients, r^2 are respectively ***0.83***, ***0.88*** and ***0.93***. (Fig.7)

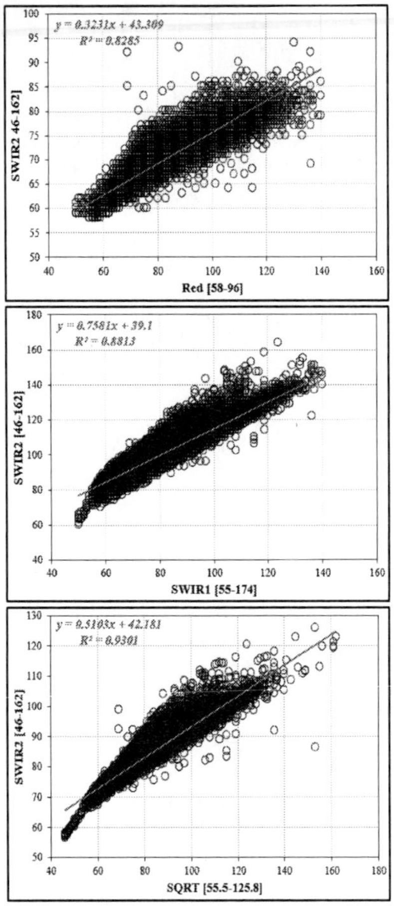

Figure 7. Linear regressions assessment

2.3.3. Technical assumptions and proposed method

The tests and approaches above have helped to formulate three technical assumptions justifying the proposed process:

i) *the SWIR2 is the main spectral band for built-up extraction with less noises*;

ii) *the square root band is the best value to extract the built-up from red and SWIR1 bands with less noise*;

iii) *the addition of the two previous, and the subtraction of the whole red and SWIR1 bands will minimize the mixed-pixel noise*.

The statements on the built-up extraction and noise reduction led to the formulation of the *Normalized Difference Built-up and Surroundings Unmixing Index*, *NDBSUI* as the following ratio:

$$\boldsymbol{NDBSUI} = \frac{(SWIR2+\sqrt{Red*SWIR1})-(Red+SWIR1)}{(SWIR2+\sqrt{Red*SWIR1})+(Red+SWIR1)} \quad (3)$$

Its efficiency in mapping the built-up was assessed using the overall accuracy and the kappa coefficient (USGS, 1990).

3. RESULTS AND DISCUSSIONS

3.1. Visual patterns and accuracy assessment

The NDBSUI map is positively stretched **[0.08-0.245]** and its visual trends match the built-up on the stacked image. Using the Support Vector Machine algorithm, the stacked image as well as the NDBSUI were classified. Then after, the stratified random sampling technique experimented by Congalton (2009) was used to assess the index accuracy displaying 300 locations. From that, the NDBSUI was proven efficient in matching the built-up area with an overall accuracy of *95.5%* and a kappa coefficient of *0.908*. (Fig.8)

When zooming into the AOIs, the index image has shown ability to clearly discriminate the others land cover categories from the built-up (Fig.9). The accuracy assessment was performed on 100 locations (pixels) for each AOI. The OA are *100%*, *96.59%* and *96.58%*, with KC of 1, *0.92* and *0.93*, respectively, in the built-up, wet soil and dry soil/rocky dominating AOIs. It gives an average OA of *97.8%*, for a KC of *0.95*.

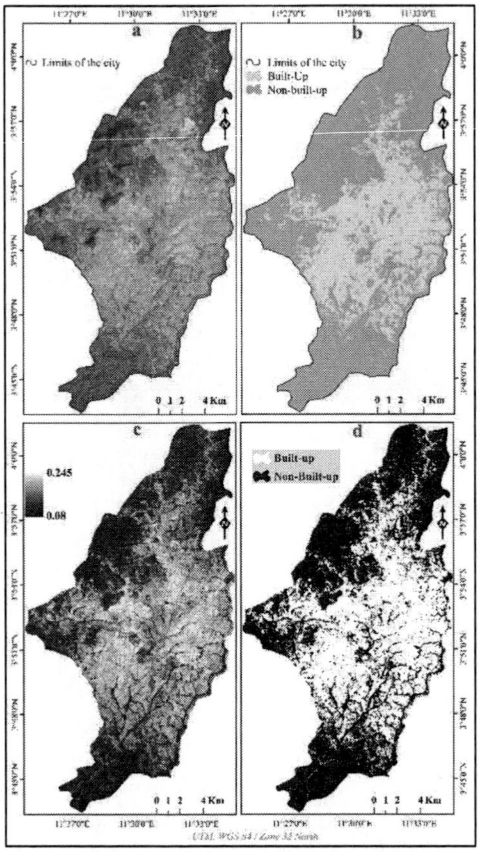

Figure 8. Visual results of the built-up extraction. *From left to right and top to down, the stacked image (a), the binary classification image highlighting built-up (b), the NDBSUI stretched* ***(c)*** *and binary views (d).*

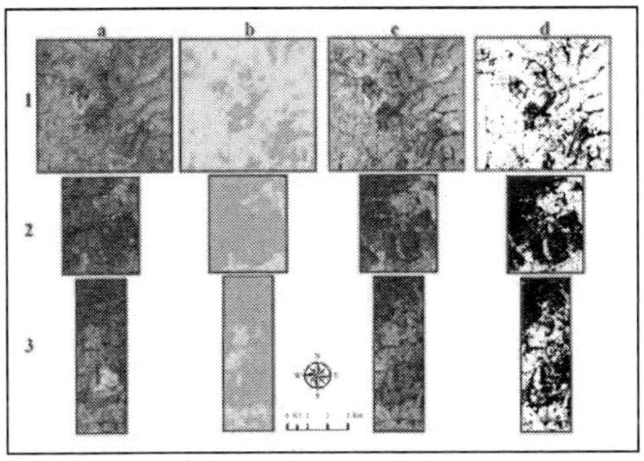

Figure 9. The NDBSUI unmixing suitability. *1, 2 and 3 represent the three AOIs while **a**, **b**, **c** and **d** are the stacked, binary classification, NDBSUI stretched and binary views.*

3.2. Comparative efficiency with existing methods

Further, the suitability of the NDBSUI in highlighting built-up features was assessed in comparison to the EBBI, NDBI and UI (Fig.10). The choice of these three indices is justified by their utilization of SWIR1 or SWIR2 as main wavelengths for built-up extraction. The built-up extraction trends are generally the same. However, from its stretching [**0.08-0.245**], the NDBSUI extract the built-up only in positive ranges, conversely to the EBBI **[-0.41817-0.659394]**, NDBI **[-0.204301-0.340909]** and UI **[0.4-0.491228]** (Fig.10, Top line). To compare their accuracies, 300 locations were also displayed using the binary images. They gave OA of *93.5%*, *92.86%* and *92.81%*, with KC of *0.87*, *0.8564* and *0.8561* respectively for EBBI, NDBI and UI (Tabl.2).

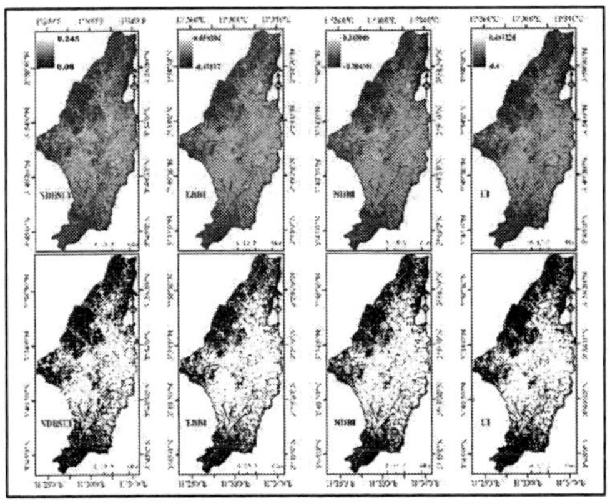

Figure 10. Indices maps. *The top line shows stretched maps and bottom binary ones.*

Index	Built-up		Non-built-up		*OA*	*KC*
	PA	***UA***	***PA***	***UA***		
NDBSUI	91	98.39	98.8	93.62	95.51	0.9076
EBBI	100	88.51	87	100	93.51	0.8701
NDBI	97.5	89.6	87.8	97.01	92.86	0.8564
UI	93.5	92.4	92	93.24	92.81	0.8561

Table 2. Compared accuracies

At larger scales of AOIs, the visual patterns show that the three others indices had helped to highlight built-up from other land cover categories, especially in the built-up and wet soil dominating AOI (Fig.11). Moreover, in the third AOI mixing built-up, dry soils, rocks and stones, the discrimination is more effective and there is low confusion than in the three other indices. Table 3 gives the details of the accuracies in the AOIs.

Index	**AOI 1**		**AOI 2**		**AOI 3**		**Average AOIs**	
	OA	*KC*	*OA*	*KC*	*OA*	*KC*	*OA*	*KC*
NDBSUI	100	1	96.5	0.92	96.8	0.93	97.8	0.95
EBBI	98.9	0.97	96.5	0.93	91.7	0.84	95.7	0.91
NDBI	96.8	0.93	94.1	0.88	95.7	0.91	95.6	0.91
UI	97.8	0.95	94.2	0.88	93.6	0.87	95.2	0.90

Table 3. Compared accuracies in the AOIs

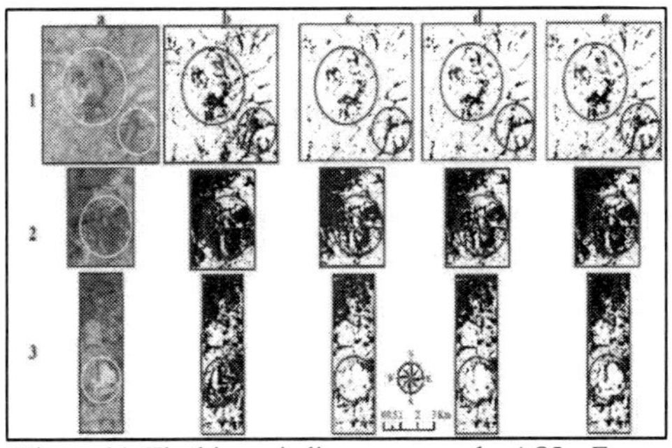

Figure 11. The binary indices maps on the AOIs. *From top to bottom, **1**, **2** and **3** represent the AOIs. From the left to the right, **a**, **b**, **c**, **d** and **e** refer to Landsat-8 OLI original image, NDBSUI, EBBI, NDBI and UI. The circles represent the areas with different detection patterns.*

The last assessment of the NDBSUI efficiency was comparing the statistics of land cover category areas. Using the binary classified image, the built-up areas extracted were compared (Tabl.4). According to the classified image, the built-up is extended over 13531.32 hectares and the other land cover categories cover 18218.97 hectares, for a total of 31750.29 hectares. The NDBSUI on its side detects built-up on 15943.95 hectares, and the non-built-up over 15806.25 hectares, for a total of 31750.2 hectares. This area is higher by 2412.63 hectares from classified image built-up area (Fig.12). When extending statistical analysis to other indices, the built-up areas of EBBI, NDBI and UI are higher by, respectively, 5035.41 hectares, 4583.34 hectares and 3328.29 hectares from the classified reference image. This confirms the efficiency of the NDBSUI to reduce the confusion between classes, better than the three other indices.

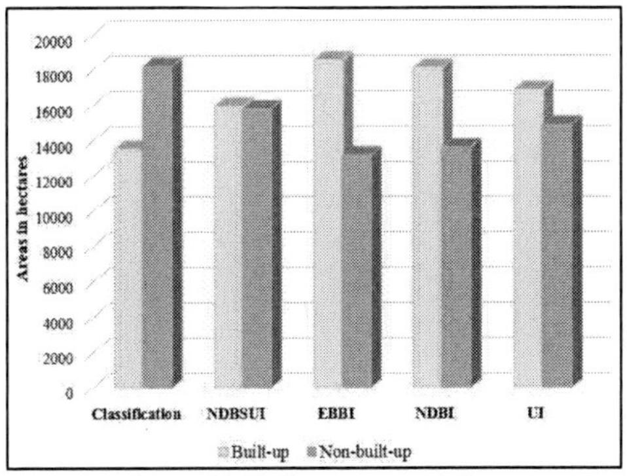

Figure 12. Built-up area per index

https://doi.org/10.5194/isprs-archives-XLIV-M-2-2020-79-2020

3.3. Test-validation on the five independents sites

The first two cities of Yokadouma and Kumba, represent the deep rainy season respectively with bimodal and monomodal rainfall (see Fig.2a), where soil is less exposed, gave the best visual as well as statistical appraisal of the built-up area. The visual patterns (fig.13) as well as the high overall accuracies and kappa coefficients testify the satisfactory application of NDBSUI (Yokadouma: OA=*98.9%*& KC=*0.88*; Kumba: OA=*97.5%* & KC=*0.947*), comparatively to EBBI, NDBI and UI.

Furthermore, the test was performed for the cities of Foumban and Ngaoundéré, representing the bio-climatic transition between the wet and the dry lands. They are respectively the validation sites for the western highlands and the high Guinea savannah, with a low forest cover and mixing of soil with the built-up. Visually, the NDBSUI has provided a better appraisal of the built-up than the other indices have (Fig.13). Statistically, the overall accuracies and kappa coefficients scores support the high performance of the NDBSUI (Foumban: OA=*97.06%* & KC=*0.918*; Ngaoundéré: OA= *95.3%*& KC=*0.89*).

The last validation site is the city of Garoua, corresponding the sub-saharian bio-climatic zone, mostly dominated by sands and transitioning from the shrubs to the desert. The NDBSUI after highlighting the built-up area according to the stacked image and the classified map had shown less confusion with the non-built-up area as well as the highest accuracy scores (OA=*74.8%* & KC=*0.42*).

The overall accuracies and kappa coefficients are recorded in Table 4.

Index	Yokadouma		Kumba		Foumban		Ngaoundéré		Garoua	
	OA	*KC*	*OA*	*KC*	*OA*	*KC*	*OA*	*KC*	*OA*	*KC*
NDBSUI	98.9%	0.88	97.5%	0.947	97.06%	0.918	95.29%	0.894	74.86%	0.421
EBBI	97.9%	0.822	92.2%	0.808	76.25%	0.457	91.2%	0.814	63.68%	0.22
NDBI	98%	0.824	97.8%	0.952	65.4%	0.30	88.3%	0.757	56.5%	0.105
UI	99%	0.918	97.1%	0.938	75.68%	0.494	91.2%	0.797	58.4%	0.024

Table 4. Validation sites accuracy

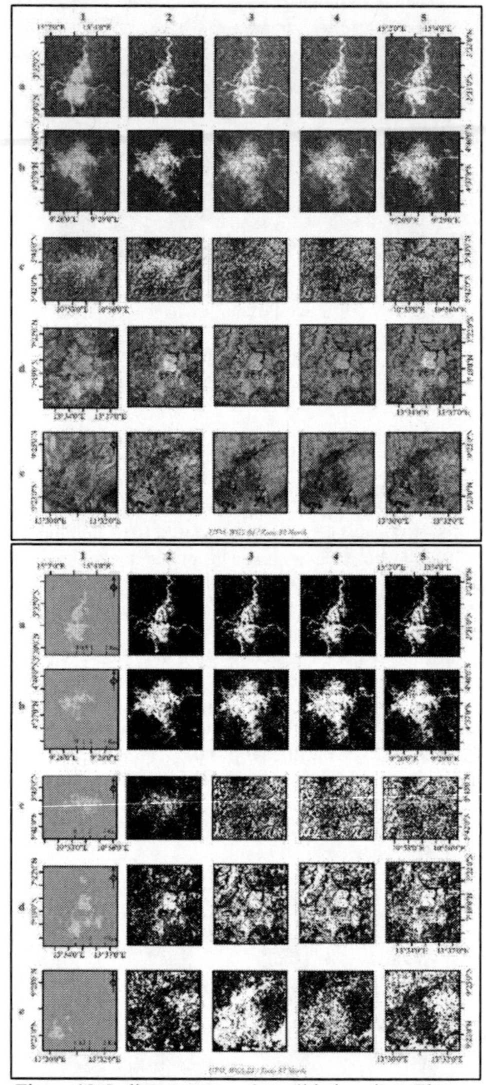

Figure 13. Indices maps on the validation sites. *The top box maps represent the Landsat8-OLI and the indices maps in stretched values. Read top-down. Yokadouma (**a**), Kumba (**b**), Foumban (**c**), Ngaoundéré (**d**) and Garoua (**e**). Read left-to-right. Landsat-OLI (**1**), NDBSUI (**2**), EBBI (**3**), NDBI (**4**) and UI (**5**). The bottom maps box represents the classification and the indices maps in binary values. Read in the same order.*

4. CONCLUSIONS

This study has addressed the problem of mixed-pixels between built-up and other surrounding features. The theory related to the red, SWIR1 and SWIR2 wavelengths in the built-up extraction, as well as reflectance analysis on the AOIs has helped to understand the mixed pixels and create the NDBSUI algorithm. The resulting accuracies were satisfactory on the experimental site of Yaoundé, as well as in the five validation sites. Despite its efficiency in extracting and unmixing built-up from surrounding features, some uncertainties related to the spatial resolution constraints that were part of challenges to

https://doi.org/10.5194/isprs-archives-XLIV-M-2-2020-79-2020 |

overcome, and the bioclimatic and ecological issues that emphasize mixed pixels extension remain to be addressed. Ongoing improvements are actually conducted to point out these noise issues from the brightness temperature and darkness approach.

AKNOWLEDGEMENTS

Authors are grateful to USGS for the satellite images freely available. They especially appreciate the support of StatsN'Maps consulting firm management, for initiating and directing this work since its presentation the ISPRS 2020 annual conference.

REFERENCES

As-syakur, A.R.; Adnyana, I.W.S.; Arthana, I.W.; Nuarsa, I.W., 2012. Enhanced Built-Up and Bareness Index (EBBI) for Mapping Built-Up and Bare Land in an Urban Area. *Remote Sens.*, 4, 2957-2970.

Bangira, T.; Alfieri, S.M.; Menenti, M.; Van Niekerk, A.; Vekerdy, Z., 2017. A Spectral Unmixing Method with Ensemble Estimation of Endmembers: Application to Flood Mapping in the Caprivi Floodplain. *Remote Sens.* 9, 1013.

Chen, J.; Li, M.; Yongxue, L.; Shen, C.; Hu, W., 2010. Extract residential areas automatically by New Built-up Index. *Proceedings of the 2010 18th International Conference on Geoinformatics*, Beijing, China, 18-20 June, IEEE.

Congalton, R.G. and Green, K. 2009. Assessing the Accuracy of Remotely Sensed Data-Principles and Practices. 2nd ed. CRC Press: Boca Raton, p.183.

Ettehadi Osgouei, P., Kaya, S., Sertel, E., Alganci, U., 2019. Separating Built-Up Areas from Bare Land in Mediterranean Cities Using Sentinel-2A Imagery. *Remote Sens.*, 11, 345.

Gonzalez, R., 2002. Digital Image Processing 3nd Edition. *Prentice Hall.* pp. 232–233.

Herold, M., Gardner, M. E. and Roberts, D. A. 2003. Spectral resolution requirements for mapping urban areas. *IEEE Transactions on Geoscience and Remote Sensing*, (41), 9, 1907-1919.

Hanqiu, X., 2010. Analysis of Impervious Surface and its Impact on Urban Heat Environment using the Normalized Difference Impervious Surface Index (NDISI). *Photogramm. Eng. Remote Sens.*, 76.

Kawamura, M., Jayamana. S., Tsujiko, Y., 1996. Relation between social and environmental conditions in Colombo Sri Lanka and urban index estimated y satellite remote sensing data. *Int. Arch. Photogram. Remote Sens.*, 31, 321-326

Kokaly, R.F., Clark, R.N., Swayze, G.A., Livo, K.E., Hoefen, T.M., Pearson, N.C., Wise, R.A., Benzel, W.M., Lowers, H.A., Driscoll, R.L., and Klein, A.J., 2017. USGS Spectral Library Version 7: *U.S. Geological Survey Data Series* 1035, p. 61.

Ngandam Mfondoum, A.H. et al. Extraction of Built-up features in complex biophysical environments by using a landsat bands ratio: results of the first tests. *Proceedings of the ASPRS 2020 annual conference virtual session*, Washington, USA, 22-26 June, p.15.

Schueler, T.R., 1994. The importance of imperviousness. Routledge: New York, USA, p.100-111.

USGS, 1990. The spatial data transfer standard. *Draft*

Rasul, A., Balzter, H., Gaylan, R., Ibrahim, F., Hameed, H.M., Wheeler, J., Adamu, B., Ibrahim, S., Najmaddin, P.M., 2018. Applying Built-Up and Bare-Soil Indices from Landsat 8 to Cities in Dry Climates. *Land.*, 7, 81.

Xu, H., 2008. A new index for delineating built-up land features in satellite imagery. *Int. J. Remote Sens.*, 29, 4269 – 4276.

Zha, Y., Gao, J., Ni, S.,2003. Use of normalized difference built-up index in automatically mapping urban areas from TM imagery. *Int. J. Remote Sens.*, 24, 583-594.

QUANTIFYING UNCERTAINTY IN CLASSIFIED POINT CLOUD DATA FOR GEOSPATIAL APPLICATIONS

Sumit Sen [1,2], Nozer Turel [2]
[1] GISE Lab, Department of Computer Science, IIT Bombay, Mumbai 400076 India
[2] Genesys International, 73-A SDF III, SEEPZ, Mumbai 400096 India sumitssen@gmail.com

KEY WORDS: Uncertainty, Classified Point Cloud, LIDAR, Segmentation, Classification.

ABSTRACT:

Classified Point Cloud data are increasingly the form of geospatial data that are used in engineering applications, smart digital twins and geospatial data infrastructure around the globe. Characterized by high positional accuracy such dense 3D datasets are often rated very highly for accuracy and reliability. However such data pose important challenges in semantic segmentation, especially in the context of Machine Learning(ML) techniques and the training data employed to provide classification codes to every point in massive point cloud datasets. These challenges are particularly significant since ML based processing of data is almost unavoidable due to the massive nature of the data that. We review different sources of uncertainty introduced by ML based classification and segmentation and outline concepts of uncertainty that is inherent in such automatically processed data. We also provide a theoretical framework for quantification of such uncertainty and argue that the standards of accuracy of such data should account for errors and omissions during auto segmentation and classification in addition to positional accuracy measures. Interestingly, the ability to quantify accuracies of ML based automation for processing such data is limited by the volume and velocity of such data.

1. INTRODUCTION

1.1 Background

Laser scanning and lidar systems have evolved as efficient techniques for capturing spatial data in a fast, efficient and highly reproducible way. It has been widely used in many fields, such as cultural heritage documentation, reverse engineering, three-dimensional (3-D) object reconstruction and digital elevation model (DEM) generation, as it can directly obtain the 3-D coordinates of objects. Classified LIDAR data has ushered a new generation of geospatial data that arguably has high locational accuracy. Highly dense point clouds therefore constitute much of the new geospatial data acquisition and are seen as the key to successful engineering projects such as corridor development or development of digital twins (Shirowzhan et al 2020). The ASPRS has helped develop several standards related to classified lidar data and in particular, the LAS file format that includes a point data record format (ASPRS, 2013). A key feature of this record is the use of a classification code that is based on table 4.9 of this standard and can have values from 0-255 which can represent classes of entities from 'Created, never classified', 'Unclassified', 'Ground' 'Building' to user defined classes. Whereas point class data can adhere to high horizontal accuracy standards as well as vertical accuracy based on the revised ASPRS standards for positional accuracy, the classification code assigned to each point data record may need attention in terms of errors of classification. Furthermore, the classification of different segments of a point cloud into multiple homogeneous regions, becomes a more complex affair when multiple points are mistakenly segmented into a region. The bigger picture of such errors in the context of a data that is held to very high standards of positional accuracy and is widely seen to be satisfying engineering grade geospatial data requirements. Furthermore, such errors are to be seen in the context of different 'automatic', 'semi-automatic' and manual approaches to classification and segmentation of lidar point cloud data.
Laser scanning techniques can be broadly classified into three categories, namely, airborne laser scanning (ALS), terrestrial laser scanning (TLS) and mobile laser scanning (MLS) or mobile lidar. Thus, there is wide heterogeneity in how lidar data is collected, processed, and the perspective each data acquisition may have.

1.2 Uncertainty, the Achilles heel of GIS

Goodchild in his seminal article of 1998 (Goodchild, 1998) discussed how the notion that spatial data could be treated by applying classical theories of measurement error, was simply too limiting. His article provided a framework to understand the issues of uncertainty in spatial data beyond positional accuracy including

- Errors: These relate to blunders, misinterpretations, misclassifications and a host of other possibilities. These range from use of wrong class codes to mixing different user defined classes in Lidar files.
- Scale problems: When local details are removed to achieve 'generalization' similar to what is done when a denser point cloud is reduced in size by reducing the point density.
- Fuzziness: Such uncertainties stem from classes and features being incompletely defined. These are very common in point cloud datasets where a large number of points are classified as 'Above Ground' or 'vegetation'
- Sampling: Sampling related uncertainties arise from missing data points which is often the case in MLS as well as ALS when features behind or underneath are not represented in the data thereby creating an uncertainty about the data points that have not been represented. Very often with point cloud datasets, partially captured features have to be discarded to ensure better visualization of the dataset.

1.3 Motivation

Maps are representations of reality (Monmonier,1977) and the era of digital twins Geospatial Augmented Reality is being used to power smart city environments (Shirowzhan et al 2020). It is also important to note the massive investment in laser scanning technologies also provide a sense of higher sophistication which lends to an assumed sense of higher reliability of the dataset. It is however to be noted that positional accuracy is only one part of the overall accuracy. This paper argues that the classification and segmentation processes used to process the point clouds account for a significant portion of the overall uncertainty and hence pushes the uncertainty attribute of such data. Given the massive number of point clouds being generated, it is imperative that the precision and recall of the Machine Learning based techniques used for both these processes are performed under significant productivity stress (Grilli et al 2017). Naturally, the false positives and false negatives of such processes aggravate the overall quality of the final product. We examine this problem both theoretically and using an example from the power utility sector.

The remaining of the paper is structured as follows. In the next section we outline the processes of segmentation and classification of lidar point clouds with reference to the challenges of maintaining high accuracy. We also discuss such challenges in terms of the training process and availability of training datasets. In the third section we discuss a theoretical framework to account for total accuracy of the data generation process that takes into consideration the current ISPRS guidelines (ASPRS, 2013) and the classification process. Finally we state our conclusions based on our observations and list some suggestions for future work in this area.

2. UNCERTAINTY IN CLASSIFICATION AND SEGMENTATION

For successful exploitation of point clouds and to better understand them, it is necessary to segment and then classify such data. The former refers to group points in subsets (normally called segments) characterized by having one or more characteristics in common (geometric, radiometric, etc.) whereas classification means the definition and assignment of points to specific classes ("labels") according to different criteria.

2.1 Accuracy of the Segmentation process

Segmentation is the process of grouping point clouds into multiple homogeneous regions with similar properties whereas classification is the step that labels these regions. Grilli et al (2017) have discussed the main categories of such segmentation into

(a) **Edge-based** Segmentation that relies on detection of borders or edges and grouping of points inside such edges

(b) **Region growing** segmentation which uses bottom-up or top-down approaches that first identifies the seed point and the growth of the segment based on color, geometrical criteria besides others.

(c) **Model fitting** based segmentation which groups points that conform to the mathematical representation of the primitive shape into a segment.

(d) **Hybrid techniques** that use a mix of above techniques.

(e) Beyond these techniques, there are **Machine Learning** based techniques that employ various clustering algorithms such as K-means and hierarchical clustering that help create the segments. The primary objective of the ML component is to cluster points based on attributes and features. This can be achieved by minimizing the sum of squares of distances between a given point and a cluster centroid; or by creating a hierarchical decomposition of a dataset
by iteratively splitting the point cloud dataset into smaller subsets based on geometrical and radiometric characteristics until each subset consists of only one object. Lu et al. (2016) presented such a hierarchical clustering algorithm which clusters any dimensional data and can be applied to mobile mapping, aerial and terrestrial point clouds.

Like the process of segmentation, there are no standard ways of evaluating the results of segmentation. Generally speaking, the uncertainty of the segmentation process can be quantified in terms of the success of segmenting all points in the cloud. A non-exhaustive list of metrics that have been used previously are listed in table 1. below.

Measure	Description
Segment-wise uncertainty	These include measure of Covariation or pairwise distances per segment identified and used extensively in medical image processing (Roy et al 2018)
Mean Voxel/Pixel wise uncertainty	Uncertainty is computed based on the number of Voxel/Pixel values that are wrongly segmented (Roy et al 2018)
Area Fit Index (AFI)	Utilizes the difference between the areas of the largest segment identified and the actual object overlapping with the segment. (Lucieer, 2004)
Average distance between a segment boundary pixel and the reference boundary	Is a distance metric that utilizes the distance of reference boundaries and the segment boundaries (Lucieer, 2004)
Closest Distance Metric (CDM)	A cost function based on similarity between boundary images (Prieto and Allen, 2003)
Percent of area lost/gained in the segments	Compares the segments to ground truth and calculates a normalized value of pixel lost/gained by segmentation (Marpu et al, 2010)

Table 1. Uncertainty measures for segmentation processes

2.2 Machine Learning process in classification of lidar point clouds

Once a point cloud has been segmented, each segment (group) of points can be labelled with a class thus to give some semantic to the segment (hence point cloud classification is often called semantic segmentation or point labelling). In the past, semantic segmentation of point clouds has mostly been investigated for laser scanner data captured from airplanes, mobile mapping systems, and autonomous robots. Some of the earliest work on point cloud classification dealt with airborne LiDAR data, with a focus on separating buildings and trees from the ground surface, and on reconstructing the buildings. Often the point cloud is converted to a regular raster heightfield, in order to apply well-known image processing algorithms like edge and texture filters for semantic segmentation (Hug and Wehr, 1997), usually in combination with maximum likelihood classification (Maas, 1999) or iterative bottom-up classification rules (Rottensteiner and Briese, 2002).

For many applications, point-cloud classification is a basic step in LiDAR processing sometimes separated from the segmentation step. Because of the complex combination of artificial and natural objects in cities, the automated classification of 3D point clouds can be a very challenging task, especially in urban areas. Broadly speaking there are three important approaches to classification of point clouds, viz, data-driven, model-driven, and hybrid (both data & model) driven. Data-driven methods, typically, use a bottom-up approach that begins with the extraction of primitives, e.g., planes, cylinders, cones, spheres or tori, followed by analyzing primitive topology in 2D or 3D space. The geometric elements of the primitives, such as lines and critical vertices of the structures, are extracted and grouped to form models. In contrast to data-driven methods, model-driven approaches involve a top-down strategy that usually begins with a hypothetical model library and then uses the point clouds to search for optimal solutions of model composition from the model library. Hybrid approaches adapt both approaches to achieve a high success rate for the classification process. Since most of the initial studies transformed the LiDAR data point cloud into image data, traditional supervised pixel-based classification techniques such as neural networks. The object-based approach relies on a user-defined hierarchical structure to classify the segmented objects, and the technique has proven to outperform the traditional pixel-based classifiers especially on airborne LiDAR data (Chen et al., 2009; El-Ashmawy, Shaker, & Yan, 2011; Minh and Hien, 2011; Sasaki et al., 2012). Various studies reported that an overall accuracy of over 80% can be achieved using object-based technique on LiDAR-derived surfaces (Yan et al, 2015). Classification algorithms from point clouds can be grouped into automatic and interactive based on the level of user interaction required. Furthermore, techniques that work on different sources of data such as (Airborne Laser scanning) considerably differ from those developed for MLS (Mobile Laser Scanning) and TLS (Terrestrial Laser Scanning), are not mutually inclusive (Yadav et al, 2018).

Techniques of classification such as Random Forests, Decision Trees (DT), Support Vector Machines (SVMs) and K-Nearest Neighbours (K-NN) have all been used for such lidar data classification tasks (Yan et al, 2015). More recently, Deep Learning (DL) and Convolutional Neural Networks (CNN) as well Reinforcement Learning (RL) have all been attempted and have shown improved results (Griffiths and Boehm, 2019). A summary of techniques of classification employed with point clouds is shown in table 2. We indicate levels of epistemic uncertainty associated with each technique as well.

It is important to note that all machine learning tasks are based on the premise of learning from a training dataset. In case of classification of lidar point clouds, such datasets are important source of Aleatoric uncertainty based on

1. Completeness (and adequacy) of the training data
2. Correctness of the training data (noise and errors)

While some of these are unavoidable uncertainties there has been much work done in the quantification of uncertainties in Machine Learning processes and categorization of what is avoidable and what is reducible. We shall examine these aspects in the next subsection.

Classification Technique	**Description**	**Level of epistemic uncertainty**
Random Forest	Used both in classification and clustering is based on collection of Decision trees	Low
Decision Trees	Used both in classification and clustering	Moderate
SVM	Support Vector Machines can be used both in classification and clustering	Moderate
K-NN	K Nearest Neighbour helps in both classification and clustering	Moderate
CNN	The commonly used platform for Deep Learning	Low-very Low
Deep RL	Recently developed platform that uses	Low-very Low

Table 2. Types of ML algorithms used for processing point cloud data and relative epistemic uncertainty reported. (Maxwell et al 2018; Griffiths, and Boehm, 2019)

2.3 Training the Machine

Machine learning research has always stressed the importance of distinguishing between (at least) two different types of uncertainty, often referred to as aleatoric and epistemic, in terms of any data generated by the 'trained machine' (Fu and Lee, 2013). Learning from data is inseparably connected with uncertainty. This is largely due to the fact that learning, understood as generalizing beyond a finite set of observed data, is necessarily based on a process of induction, i.e., replacing specific observations by general models of the data-generating process. Thus, a training dataset helps the learning of clustering patterns, decision trees, or CNN and then such a 'learnt modeling' is applied to a large data to complete segmentation and/or classification tasks. Naturally, such models are never provably correct but only hypothetical and therefore uncertain, and the same holds true for the predictions produced by a model. In addition to the uncertainty inherent in inductive inference, other sources of uncertainty exist, including incorrect model assumptions and noisy or imprecise data.

It is desirable and should be considered as a key feature of any machine learning method that is employed in tasks of high precession. While there are significant investments to raise the positional accuracy of point cloud data, it is important to invest in quantifying and reducing uncertainty of the ML derived processes of segmentation and classification.

It is important to understand the challenges of distinguishing between two very different sources of uncertainty: aleatoric uncertainty, which is due to statistical variability and effects that are inherently random, and epistemic uncertainty which is caused by a lack of knowledge(Kruse et al, 1991). Furthermore, classification using ML can be seen as a decision making problem and classification problem is often formalized within the framework of Bayesian decision theory. Detailed models have thus been developed through bayesian inference that

https://doi.org/10.5194/isprs-archives-XLIV-M-2-2020-87-2020 | © Authors 2020. CC BY 4.0 License.

enable to quantify the uncertainty (Senge et al, 2014). Furthermore, it has been reported that total variance in machine learning processes can

indeed be decomposed as $\sigma_{combined}=\sigma^2_{aleatoric}+\sigma^2_{epistemic}$ (Clements et al 2019)

Thus, we recognize that the combined uncertainty of the machine learning based classification outputs is explained by

$$\psi_{combined}=\psi_{aleatoric}\cdot\psi_{epistemic} \quad (1)$$

where $\psi_{epistemic}$ depends on the choice of the model, whereas $\psi_{aleatoric}$ is based on the nature and extent of the training data used. We use this understanding to develop an overall account of uncertainty for classified point cloud datasets.

THEORETICAL ACCOUNT OF UNCERTAINTY IN GENERATING CLASSIFIED POINT CLOUDS

While the different sources of uncertainty in lidar point clouds beyond positional uncertainty have been discussed in the context of the segmentation and classification process, we now focus on developing a combined framework of uncertainty by attempting to quantify uncertainty by developing indicators of uncertainty at each step of the development of a classified point cloud.

2.4 RMSE and uncertainty values based on positional accuracy

The accuracy of the position of features, including horizontal and vertical positions, with respect to horizontal and vertical datums. It is usually represented as Root Mean Square Error and is simply derived from the sum of the squares of the difference in the coordinates of the data points and the corresponding ground-truth. is expressed as

$$RMSE=\sqrt{[\frac{\sum((E_{data,i}-E_{true,i})^2)}{n}]} \quad - (2)$$

Thus, horizontal RMSE is expressed as

$$RMSE_{Hor}=\sqrt{[\frac{\sum((RMSE_E)^2+(RMSE_N)^2)}{n}]} \quad - (3)$$

Similarly vertical RMSE is expressed as

$$RMSE_{Ver}=\sqrt{[\frac{\sum(RMSE_Z)^2}{n}]} \quad - (4)$$

These measures have been incorporated into various standards including the ASPRS guidelines (ASPRS, 2015), FGDC standards (www.fgdc.gov) besides others. Very often data collection projects specify minimum requirements for horizontal and vertical accuracy. However, it is highly likely that collected is collected at different levels of RMSE, especially when the point clouds from aerfgdcial lidar surveys and mobile or terrestrial lidars are combined. Such spatially varying uncertainties pose a unique challenge for quantifying uncertainty in the positional accuracy of a combined dataset. For example, the RMSE values of a combined dataset may be well below the ASPRS threshold (ASPRS, 2015) but a certain small section (probably surveyed using mobile lidar) is very high. It is therefore pragmatic to use localized RMSE threshold values such as

$$RMSE_{local}=\sqrt{[\frac{\sum_{LocalArea}(E_{data,i}-E_{true,i})^2}{n}]} \quad - (5)$$

2.5 Accounting for uncertainty introduced in segmentation

Uncertainties introduced in the segmentation step include the exclusion of certain points from a segment or inclusion of unnecessary points. While it is not necessary to drop any points from the dataset, at this stage, the inability to group points from a segment together reduces the ability to classify the wrongly segmented points together with the object in our experience. Sparse points or highly dense datasets suffer from such challenges and hence become too 'noisy' for classification tasks.

While the parameters that can represent accuracy of the segmentation process have been discussed earlier, we recollect that CDM, AFI as well as the percent of area lost/gained in segments are all normalized measures and can be helpful in quantifying the increase in uncertainty. We can express this increase in uncertainty as

$$\psi_{Seg}=\frac{1}{CDM_{normalized}} \quad - (6)$$

2.6 Uncertainty due to machine learning processes

The machine learning process accounts for at least two categories of uncertainty that is introduced. If the segmentation step utilizes a ML based clustering algorithm, the increase in uncertainty of equation 6 could be subdivided into epistemic and aleatoric uncertainties (Hüllermeier and Waegeman, 2019). The characterization of such uncertainty is thus the same as that for the classification step discussed henceforward.

Epistemic uncertainty of the ML based classification is primarily dependent on the suitability of the ML algorithm for the learning task. As expressed in table 2, these are characteristic of each ML technique although the techniques of calculating such values have only evolved recently. Since the highest possible values of accuracy in machine learning

While 'class balanced datasets', clear distinction between different classes and possibly with cases with fewer number of classes are expected to yield better results in aleatoric sense (Fu and Lee, 2013). Thus, it is possible to obtain more specific components of aleatoric uncertainty (i) based on coverage (and hence volume) and quality of the training dataset and (ii) the nature of the classes that are used for the ML (how distinct are the features that determine membership to a class). It is also important to note that higher uncertainty in segmentation (equation 6) is expected to result in higher aleatoric uncertainty and vice-versa. Thus, we claim $\psi_{aleatoric}\propto\psi_{Seg}$

Also if χ , Ω represent the size of the training data and the quality respectively on a normalized scale (where 0 represents the best case scenario and 1 is the worst), then

$$\psi_{aleatoric}=K\cdot\chi\cdot\Omega\,\psi_{Seg} \quad - (7)$$

where K is otherwise constant.

This contribution has been peer-reviewed.
https://doi.org/10.5194/isprs-archives-XLIV-M-2-2020-87-2020 | © Authors 2020. CC BY 4.0 License.

2.7 Other sources of uncertainty

Many other sources of uncertainty are introduced in the context of point cloud data which include imprecision of the geographic referencing, incompleteness of the scanning whereby certain parts or segments may be missing or partial, fuzzy nature of boundaries of the segments and objects being captured (and later classified). We denote such uncertainty as ψ_{multiple} and postulate that this can be further decomposed into measurable components.

2.8 Combined measures of Uncertainty

It is now possible to combine the multiple sources of uncertainty to account for an overall uncertainty value for the processed lidar point cloud. It is possible to add all uncertainties including RMSE and the different ψ values. It is also possible to use weightages for each. However, it may also be pragmatic to use the different values without combining them as this can help identify the source of uncertainty and help reduce it.

2.9 Examining uncertainty in power utility datasets

Power utilities have turned to using lidar point clouds to manage their assets and ensure preventive maintenance. We use data obtained from one such project to highlight the aspect of uncertainty in lidar point clouds as discussed earlier.

With high positional accuracy, the data shows very high positional accuracy in general ($RMSE_{Hor}$<0.05 ft, $RMSE_{ver}$<0.05 ft). However, at segmentation level we notice some level of failures that causes noise (see figure 1). Note that the noise points are left alone by the classifier. If the failure occurs at classification level (see figure 2), the whole segment is wrongly labeled as a different class (in this case a vertical linear object is classified as a pole). In figure 3 we notice that the dense point cloud above the cables is classified as vegetation and it is possibly an epistemic uncertainty associated with the classifier. In some cases segmentation fails to identify the cable and hence classification fails to label them correctly. The semantic segmentation hence shows an uncertainty value >0 and can range up to 0.2 (beyond which the classification is reattempted). The ability to locate the power cable in a continuous manner is critical for the application of this lidar point cloud and as seen in figure 4, the failure to have a continuous cable could result from uncertainty in the data capture, segmentation or the classification. These have all been accounted for in our theoretical discussion.

Figure 1. Salt and pepper noise in point cloud data resulting from poor segmentation.

Figure 2. Misclassification of a linear object as a pole. It can be argued that the training data did not have any such vertical feature to be classified and hence the misclassification is a result of aleatoric uncertainty.

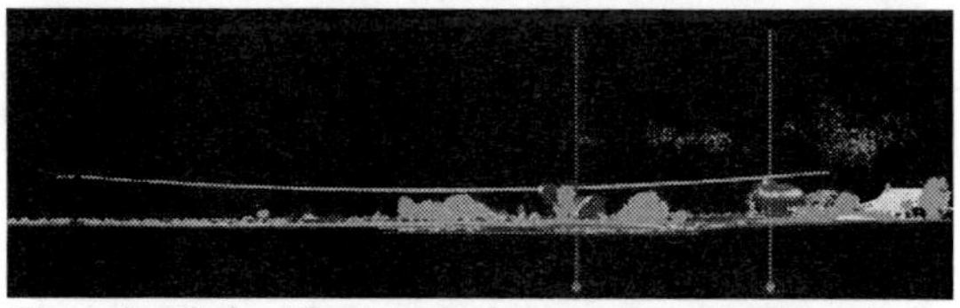

Figure 3. Misclassification of a dense segment of noise points above the cables as vegetation. Since the points are dense and hence would have been labelled by the classification algorithm in any case. Thus, it is easy to suspect epistemic uncertainty in this case.

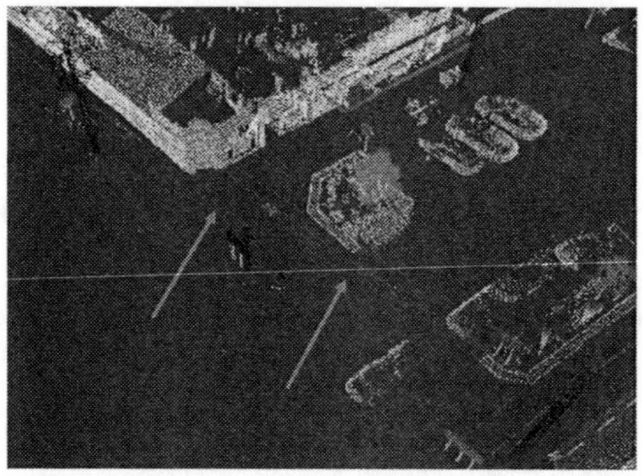

Figure 4. Failure to classify cables could be caused because of the uncertainty in the collecting data (incomplete capture), faulty segmentation as well as classification errors.

2.9.1 Computing uncertainty and reclassification

It is important to note that the cost of creating classified point cloud data includes a major component of segmentation and classification. As the cost of lidar surveys outpace that of traditional or photogramatery based surveys, it is important to remember that even with very low RMSE values (<0.05ft), there are significant uncertainties that remain associated with lidar point clouds. Due to the dense point clouds and massive datasets, it is convenient (almost compulsory) to use ML based techniques to generate classified point clouds. The cost of processing data in terms of manually validating classified power utility data for the example above can exceed one person-day for less than 10 km. Furthermore, in worst cases the classification is required to be repeated and hence cause higher expenses and project delays.

Since manual classification forms the basis of most training datasets as well as validation of the classified point clouds, human errors and omissions are critical to the framework of point cloud data processing. It is important to account for all investments in creating the cloud point data while stating its quality and not relying only on positional accuracy.

https://doi.org/10.5194/isprs-archives-XLIV-M-2-2020-87-2020

The volume and velocity of point cloud data processing by what is commonly termed as spatial big data analysis (Shirowzhan et al 2019) makes the quantification of uncertainty a challenging task (and an expensive one). However, it is important to ensure that the data adheres to acceptable quality standards as shown in the running example.

3. CONCLUSIONS AND FUTURE WORK

We have reviewed multiple sources of uncertainty in lidar point clouds and have shown that a holistic approach to spatial data quality is necessary. The specific need to account for uncertainty introduced at the segmentation and classification stage has been highlighted. The understanding of epistemic and aleatoric uncertainties to reduce classification errors in the machine learning based classification of point clouds is an important aspect of this holistic approach.

Our experience with classification shows that there multiple aspects of uncertainty associated with the acquisition and processing of lidar point clouds. With higher precision of the laser based sensing mechanism and prescribed standards of accuracy, it is important to recognize that manual 3D segmentation is a complex procedure which requires a skilled user, patience and an acute eye for detail. Although there are multiple available training datasets, these are not necessarily contextual geospatial applications at hand. We demonstrate through the power utility example that the cost of using lidar point data is closely related to the uncertainties associated with the uncertainties and hence to overall spatial data quality. With higher investments in lidar based spatial data, it is therefore important to revisit Goodchild's article (Goodchild, 1998), and work towards specification of uncertainties endemic to generation of classified point cloud data.

We believe that this area of spatial data quality is an emerging topic of research. Some of the areas for future exploration (but not limited to) are:

- Develop a detailed framework to statistically account for uncertainties in the machine learning framework, especially in deep learning. Such work can benefit from the progress made by the data science communities in providing epistemic and aleatoric values of uncertainty. Of particular interest are the epistemic uncertainties of a classifier, especially using Monte Carlo Dropout (MCD) (Miok et al 2019).
- More importantly, it is important to develop metrics for spatial data quality that considers the different sources of uncertainty rather than positional accuracy alone. Thus, the theoretical framework stated in this paper could be the basis of a spatial data quality metric and could be used to understand the cost of lidar data projects (Hummel et al 2011)
- Training datasets used to classify point cloud data are often not inline with classification requirements. Thus, it is important to understand the role of inadequate training in the quality of classified point datasets and hence will be useful to compare the relative uncertainty of data produced by classifiers that have been trained using different datasets (Fu and Lee, 2013; Wang, et al 2019).
- The aspects of semantics in 'semantic segmentation' and the use of shared vocabularies are important to resolve ambiguities in 'user defined' classes of lidar point clouds. Research related to geospatial ontologies and probabilistic ontologies (Sen, 2008) can provide solutions to resolve such challenges.

ACKNOWLEDGEMENTS

We acknowledge the support of Genesys International and colleagues for the use case and data related to the power utility example.

REFERENCES

ASPRS, 2015. ASPRS positional accuracy standards for digital geospatial data. Photogrammetric Engineering and Remote Sensing, 81(3), pp.1-26.

ASPRS, 2013. LAS specification, version 1.4 – R13 (PDF). American Society for Photogrammetry and Remote Sensing. 2013-07-15 available at https://www.asprs.org/wp-content/uploads/2010/12/LAS_1_4_r13.pdf

Chen, Y., Su, W., Li, J., Sun, Z., 2009. Hierarchical object oriented classification using very high resolution imagery and LiDAR data over urban areas. Advances in Space Research 43 (7), 1101–1110.

Clements, W.R., Robaglia, B.M., Van Delft, B., Slaoui, R.B. and Toth, S., 2019. Estimating risk and uncertainty in deep reinforcement learning. arXiv preprint arXiv:1905.09638.

Clements, W.R., Robaglia, B.M., Van Delft, B., Slaoui, R.B. and Toth, S., 2019. Estimating risk and uncertainty in deep reinforcement learning. arXiv preprint arXiv:1905.09638.

El-Ashmawy, N., Shaker, A., , Yan, W. Y., 2011. Pixel vs object-based image classification techniques for LiDAR intensity data. International Archives of the Photogrammetry, Remote Sensing and Spatial Information Sciences 38 (Part 5/W12), 43–48.

Fu, J. and Lee, S., 2013. Certainty-based active learning for sampling imbalanced datasets. Neurocomputing, 119, pp.350-358.

Goodchild, M.F. "Uncertainty: the achilles heel of gis?" Geo Info Systems, vol. 8, no. 11, pp. 50–52, 1998.

Griffiths, D. and Boehm, J., 2019. A review on deep learning techniques for 3D sensed data classification. Remote Sensing, 11(12), p.1499.

Grilli, E., Menna, F. and Remondino, F., 2017. A review of point clouds segmentation and classification algorithms. The International Archives of Photogrammetry, Remote Sensing and Spatial Information Sciences, 42, p.339.f

Hug, C. and Wehr, A., 1997. Detecting and identifying topographic objects in imaging laser altimeter data. International archives of photogrammetry and remote sensing, 32(3 SECT 4W2), pp.19-26.

Hüllermeier, E. and Waegeman, W., 2019. Aleatoric and epistemic uncertainty in machine learning: A tutorial introduction. arXiv preprint arXiv:1910.09457.

Hummel, S., Hudak, A.T., Uebler, E.H., Falkowski, M.J. and Megown, K.A., 2011. A comparison of accuracy and cost of LiDAR versus stand exam data for landscape management on the Malheur National Forest. Journal of forestry, 109(5), pp.267-273.

Kruse, R., Schwecke, E. and Heinsohn, J., 2012. Uncertainty and vagueness in knowledge based systems: numerical methods. Springer Science & Business Media.

Lu, X., Yao, J., Tu, J., Li, K., Li, L. and Liu, Y., 2016. Pairwise Linkage for POint Cloud Segmentation. ISPRS Annals of Photogrammetry, Remote Sensing & Spatial Information Sciences, 3(3).

Lucieer, A., 2004. Uncertainties in Segmentation and their Visualisation (Doctoral dissertation, Utrecht University and International Institute for Geo-Information Science and Earth Observation (ITC)).

Maas, H.G., 1999, June. The potential of height texture measures for the segmentation of airborne laser scanner data. In Fourth international airborne remote sensing conference and exhibition/21st Canadian symposium on remote sensing (Vol. 1, pp. 154-161).

Marpu, P.R., Neubert, M., Herold, H. and Niemeyer, I., 2010. Enhanced evaluation of image segmentation results. Journal of spatial science, 55(1), pp.55-68.

Maxwell, A.E., Warner, T.A. and Fang, F., 2018. Implementation of machine-learning classification in remote sensing: An applied review. International Journal of Remote Sensing, 39(9), pp.2784-2817.

Minh, N. Q., Hien, L. P., 2011. Land cover classification using LiDAR intensity data and neural network. Journal of the Korean Society of Surveying, Geodesy, Photogrammetry and Cartography 29 (4), 429–438.

Miok, K., Nguyen-Doan, D., Zaharie, D. and Robnik-Šikonja, M., 2019, September. Generating Data using Monte Carlo Dropout. In 2019 IEEE 15th International Conference on Intelligent Computer Communication and Processing (ICCP) (pp. 509-515). IEEE.

Monmonier, M.S., 1977. Maps, Distortion, and Meaning. Resource Paper No. 75-4.

Mosavi, A., Ghamisi, P., Faghan, Y. and Duan, P., 2020. Comprehensive Review of Deep Reinforcement Learning Methods and Applications in Economics. arXiv preprint arXiv:2004.01509.

Mosavi, A., Ghamisi, P., Faghan, Y. and Duan, P., 2020. Comprehensive Review of Deep Reinforcement Learning Methods and Applications in Economics. arXiv preprint arXiv:2004.01509.

Prieto, M.S. and Allen, A.R., 2003. A similarity metric for edge images. IEEE Transactions on Pattern Analysis and Machine Intelligence, 25(10), pp.1265-1273.

Rottensteiner, F. and Briese, C., 2002, September. A new method for building extraction in urban areas from high-resolution LIDAR data. In International Archives of Photogrammetry Remote Sensing and Spatial Information Sciences (Vol. 34, No. 3/A, pp. 295-301). Natural Resources Canada.

Roy, A.G., Conjeti, S., Navab, N. and Wachinger, C., 2018, September. Inherent brain segmentation quality control from fully convnet monte carlo sampling. In International Conference on Medical Image Computing and Computer-Assisted Intervention (pp. 664-672). Springer, Cham.

Sasaki, T., Imanishi, J., Ioki, K., Morimoto, Y., Kitada, K., 2012. Object-based classification of land cover and tree species by integrating airborne LiDAR and high spatial resolution imagery data. Landscape and Ecological Engineering 8 (2), 157–171.

Sen, S., 2008. Framework for probabilistic geospatial ontologies. International Journal of Geographical Information Science, 22(7), pp.825-846.

Senge, R., Bösner, S., Dembczyński, K., Haasenritter, J., Hirsch, O., Donner-Banzhoff, N. and Hüllermeier, E., 2014. Reliable classification: Learning classifiers that distinguish aleatoric and epistemic uncertainty. Information Sciences, 255, pp.16-29.

Senge, R., Bösner, S., Dembczyński, K., Haasenritter, J., Hirsch, O., Donner-Banzhoff, N. and Hüllermeier, E., 2014. Reliable classification: Learning classifiers that distinguish aleatoric and epistemic uncertainty. Information Sciences, 255, pp.16-29.

Shirowzhan, S., Tan, W. and Sepasgozar, S.M., 2020. Digital Twin and CyberGIS for Improving Connectivity and Measuring the Impact of Infrastructure Construction Planning in Smart Cities.

Wang, G., Li, W., Aertsen, M., Deprest, J., Ourselin, S. and Vercauteren, T., 2019. Aleatoric uncertainty estimation with test-time augmentation for medical image segmentation with convolutional neural networks. Neurocomputing, 338, pp.34-45.

Wang, G., Li, W., Aertsen, M., Deprest, J., Ourselin, S. and Vercauteren, T., 2019. Aleatoric uncertainty estimation with test-time augmentation for medical image segmentation with convolutional neural networks. Neurocomputing, 338, pp.34-45.

Yadav, M., Lohani, B. and Singh, A.K., 2018. Road surface detection from mobile lidar data. ISPRS Annals of the Photogrammetry, Remote Sensing and Spatial Information Sciences, 4, pp.95-101.

Yan, W.Y., Shaker, A. and El-Ashmawy, N., 2015. Urban land cover classification using airborne LiDAR data: A review. Remote Sensing of Environment, 158, pp.295-310.

This contribution has been peer-reviewed.
https://doi.org/10.5194/isprs-archives-XLIV-M-2-2020-87-2020 | © Authors 2020. CC BY 4.0 License.

PHOTOGRAMMETRIC MODELING OF SUBTERRANEAN FEATURES THROUGH THREE-DIMENSIONAL SOFTWARE ANALYSIS

J. Sepulveda [1,2,*], J. Capps [1], K. Johnson [1], C. Parada [1], A. Garcia[1], K. Sestak [1]

[1] Dept. of Geography, Texas A&M University, 797 Lamar St, College Station, TX - (jaroncapps, auggie_97, kallidj15, chelsea.parada, jorgesepulveda117, k.sestak01)@tamu.edu
[2] Dept. of Computer Science and Engineering, Texas A&M University, 710 Ross St, College Station, TX - jorgesepulveda117@tamu.edu

KEY WORDS: Photogrammetry, Subterranean Environments, Structure from Motion, Karst, Unity, Cave Conservation, Point Clouds

ABSTRACT:

LiDAR is a popular and accurate method for mapping that can be utilized for three-dimensional model analysis. However, the equipment set-up and usage can become tedious, and ultimately impractical when applied to locations that are remote and confined in nature. In this investigation, three-dimensional analysis was conducted within a cave system. With this, limitations of LiDAR technology in these conditions become prominent; mapping non-planar surfaces can cause a potential decrease of the quality of the point cloud data. In all, a LiDAR application would be an inefficient use of methodology to conduct this investigation. This prompted a need to set-up and conduct a photogrammetric based evaluation. With this, smartphone camera technology was used in conjunction with free-to-use software and three-dimensional modeling applications. Through the use of photogrammetric concepts and structure from motion software, a three-dimensional model of the cave can be generated. Long term, this model can also be utilized to document the impact and health of the cave system. For the methodology, the on-sight portion of the investigation relied heavily on smartphone camera technology. The procedure draws parallels to drone paths; specifically, two flight-plans were developed to evaluate different perspectives within a 15 by 15 meter space in the cave. Within each flight path, the use of photo overlapping techniques established a denser and more fluid point cloud model. Once the data was processed, two different three-dimensional models of the cave were created. From those models, the point cloud data was extracted in order to merge the two separate models. Afterwards, the models underwent several format conversions in order to import it into the Unity game engine. The final result is an accurate three-dimensional model of the cave that is viewable and playable in a simple video game platform.

1. INTRODUCTION

Karst topography is made up of complex systems that are shaped by fluvial forces and can act as conduits for groundwater. They can form from any soluble rock, the most common being limestone. The solution of limestone is a reaction of the groundwater, which is naturally acidic from carbon dioxide dissolving away the calcium carbonate where limestone originates. Resulting in geological features including, but are not limited to, disappearing streams, sinkholes, and caves. These features have become points of research interest among several hydrogeologists due to groundwater movement being unusually fast within karst. Caves, in particular, are a source of recreation for many in the public and has resulted in maps of thousands of caves worldwide. However, some of these maps may not be ideal in conveying minute details and spatial references. The documentation of these features realizes a lack of cited ecosystem exchanges and reveals the significance of global contributions.

* Corresponding author

To understand planetary interactions of these features, particularly caves, more detailed research in their mapping is required. It is important to understand these environments to understand both the biota that rely on them and water resources upon which we rely on. With this, a fascination to find a medium to accurately map or chart a karst feature led to the inference that a mapping software or modeling application would be needed for a deeper investigation. This prompted an interest towards both LiDAR (Light Detection and Ranging) technology and photogrammetric 3D modeling.

1.1 Potential Applications

With LiDAR, modeling karst features and establishing algorithms to calculate precise measurements and distances would aid in the illustratration of subterranean interactions. However, many variables must be accounted for before the actions of an investigation of this nature are executed. To start off, within a karst system, edges, non-planar, and variable depth of surfaces can result in low point cloud concentrations (Gurram, P., Lach, S., Saber, E., Rhody, H., & Kerekes, J., 2007). Point clouds are a series of data points that have an X,Y and Z location all stored in a file. This would lead to a decline in the quality of data. The narrow corridors and passageways within a cave can also become an issue when attempting to collect data by being physically impassable or difficult to navigate with delicate equipment.. This prompts a safety and

https://doi.org/10.5194/isprs-archives-XLIV-M-2-2020-95-2020

mobility concern due to the combination of equipment care and potential tight spaces. These factors, combined with cost and limited knowledge on how to use LiDAR technology for cave mapping is why more traditional survey methods with suuntos and Disto-Xs are common and preferred ways of mapping among the caving community. Therefore, from the constraints that both LiDAR and traditional cave mapping have, it was reasoned that the use of a more portable, cheaper method to LiDAR would be utilized in the involvement towards the creation of an accurate 3D model. With this in mind, a potential solution is the use of smartphone camera technology, personal measurements, and modelling applications. This will not only document the cave with photography but the resulting 3D map will provide as a tool of safe passage for cavers and researchers alike. The data collected is in the form of in-situ measurements, or on-sight data collection, establishing and emphasising the factor of human observations and interactions within a karst feature.

1.2 Hypothesis

This investigation explores the practicality of accurately mapping and modelling a cave environment using smartphone camera technology processed with three-dimensional model analysis. This is to explore a cost-effective and lighter load bearing solution compared to LiDAR. This idea promotes a portable experience, allowing for more data to be collected in tight spaces within a cave system, and establishing preventive measures to be taken during data collection to limit ecological disturbances. If the combination of smartphone camera technology with the use of three-dimensional photography stitching is a medium that can bypass the use of LiDAR equipment within a karst feature, then cave systems can be accurately charted and mapped to help determine ecological health and natural changes, and promotes further interest of research for remote environments.

2. METHODS

The investigation took place within Inner Space Caverns in Georgetown, Texas. This cave offered the opportunity to collect data in a large room on the eastern side of the cave. The room simulated different experiences for surveyors, which established different methods and strategies to sample and collect the data. All in all, this location illustrated the importance of portable data collection due to the array of topological changes and limited mobility.

2.1 Smartphone Camera Application

During the on-sight portion of the investigation, pictures of the cave were captured with an iPhone XS. The investigation was centralized on two flight-paths within an area of 225 square meters. The two flight-paths evaluated different point-of-views of the cave, establishing a perpendicular contrast between the two. During the first flightpath, the collection of photographs started at one corner, and subsequent pictures were taken along a laid-out line at one-meter intervals for 15 meters. After completing a line, the camera was rotated, capturing three photographs. One at the last point of the line, and two after rotating counter-clockwise at 90 and 180 degrees. After the collection of the photograph at 90 degrees, one step was taken before the collection of the next photograph, prompting the start of a new line. This provided a good amount of overlap between the pictures for the entire cave room, allowing for an increase of quality for the development of a 3-D model. Due to this, the establishment of the minimum requirement of 83% overlap between the taken photographs was achieved, prompting a similar technique to the construction of drone flight-paths. This process continued until the entire marked-off area was captured, resulting in the completion of the first flightpath. During the second flightpath, the adjacent corner from the first flightpath became the new starting point. This approach was similar to the first; however, this application allowed for a different view of the cave, ultimately increasing the quality of the point cloud for modelling purposes.

2.2 3-D Model Processing

After the data collection, 500 pictures of the cave room were collected. 258 of the pictures faced the front side of the cave and the remaining 242 faced the rear. The downloaded pictures of the cave were filtered according to their respective side of the cave. This prompted the 3-D modeling software to avoid high amounts of uncalibrated pictures while performing the modeling tasks. One set of images focused on the front side of the cave, while the other set focused on the rear side of the cave. The area of the cave that was photographed is marked by a blue square seen below in Figure 1.

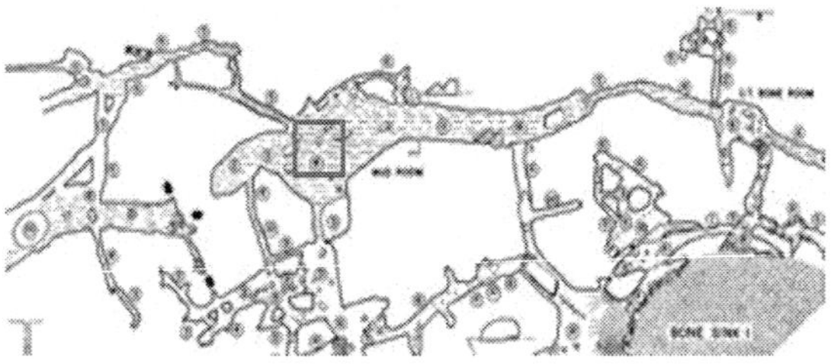

Figure 1. A planar view of the region of Inner Space Cavern where the experiment was conducted.

With the pictures filtered, two models were made using Pix4DMapper©, a proprietary 3-D modeling software that uses photogrammetry and computer vision algorithms to create said models (Pix4Dmapper Development Team, 2019). The previously mentioned software provided two 3-D models of both sides of the cave area alongside the point cloud data in a LAS file format. The LAS format is a commonly used file format for LiDAR point cloud data. Afterwards, the two different 3-D models needed to be merged into one so a model of the entire cave would be available.

To merge the separate 3-D models into one, LAStools was used. LAStools is a command line toolkit written in the C++ programming language developed by rapidlasso GmbH. It was used for processing and editing point cloud data. It can do multiple tasks such as creating Digital Elevation Models (DEM), establishing point cloud file attributes and, most importantly, merge multiple point clouds into one (Rapidlasso Development Team, 2019). Using this toolkit, the point clouds were successfully merged into a single data set. The resulting point cloud can be seen in Figure 2.

https://doi.org/10.5194/isprs-archives-XLIV-M-2-2020-95-2020 | © Authors 2020. CC BY 4.0 License.

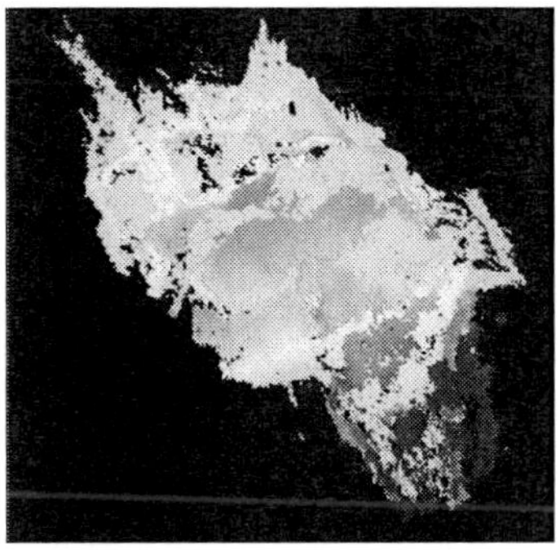

Figure 2. Top down view of the Cave's merged point cloud made from two existing point clouds, respectively the rear view and front view of the pictures. This is shown using lidarview.com.

2.3 Unity Game Engine

One of the main goals for this investigation is to import the 3-D model of the cave into a video game, specifically the Unity game engine. Game engines are developmental kits that contain libraries and components that are used in the creation of video game platforms. With this, Unity is a popular game engine made by Unity Technologies in 2005. Since then, it has grown to be a commonly used engine in modern video games. It's user friendly, hands-on tutorials establishes simple notions to help create a platform that can aid in the navigation and manipulation of data (Unity Technologies, 2019).

Now that a merged point cloud model was available, it needed to be converted to a format that Unity can recognize. LAS files are not easily convertible into game engines and the cave model had to be reformatted. Fortunately, Keijiro Takahashi created a Unity module called Pcx. This module allows for 3-D models to be imported into a Unity game with ease (Takahashi, 2019). However, the module only reads Polygon File Format files, which is a different format for 3-D models but with polygons in-between the points.

To achieve this, CloudCompare was used to convert the cave's LAS file into a PLY format. CloudCompare is an open source 3-D point cloud processing software (CloudCompare Development Team, 2019). Similar to LAStools, CloudCompare can process point cloud data but it can also provide PLY format files. The software has an easy-to-use command line interface which was used to convert the point cloud model into the desired format.

After CloudCompare provided a PLY file, the Pcx module easily imported the cave into Unity. In Unity the cave was properly oriented. An invisible floor was placed in the cave and a movable first person camera was added into the game. Now, the photographed cave area was brought into a simple video game, where players can look around the cave and move around in it. There is a satisfactory level of detail in the cave. The point cloud model provided an accurate amount of the cave's detail in the final result. In Figure 3, the first person camera is facing the entrance of the front side of the cave.

Figure 3. The finalized product of the merged point clouds of the cave in Unity.

3. RESULTS

3-D Models were made from the two flight paths. Point cloud models were then extracted from the models and merged into one single point cloud. The resulting point cloud is illustrated in Figure 2. From the 500 pictures that were taken, 450 were successfully processed into the finalized 3-D model. The remaining 50 were deemed as uncalibrated by Pix4DMapper©.

The combined model was then converted into a different format that is compatible with the Unity game engine. This allows users to be able to view and interact with the cave from a first-person perspective. The game also allows the user to walk around the mapped cave area and observe its interior.

4. CONCLUSION

Many obstacles were faced during the investigation, with one being the topography. The enclosure's topography varied as the pictures were taken, which made it difficult for the photographer to accurately capture the image alongside the proposed flight path.

Next, due to overlap, the resulting model of the mapped area produced certain regions to be more defined. On point cloud, the overlaps were more concentrated, which allowed higher quality data. This established a flow between flight paths that created a better model resolution. Our modeling system didn't provide the exact measurements of the overlap, but it is estimated to be over 80%. This was due to our basic understanding of photogrammetry which let to tight flight paths.

Also, the angle of the phone had to be adjusted due to poor lighting and casting of shadows. Each picture was uniquely altered due to the various corners, edges, and sides shown throughout the cave.

Additionally, some software issues were found along the way. Pix4Dmapper wasn't able to create a merged 3-D model so that prompted looking into other software solutions such as LASTools and CloudCompare.

Furthermore, there is no existing LiDAR data to be able to compare the results from the created 3-D model. There is an assumption that LiDAR is more accurate but the only way to determine this is to map the cave with the aforementioned equipment.

Overall, a representation of the cave was still successfully produced. Though it may not be as detailed as LiDAR, this

This contribution has been peer-reviewed.
https://doi.org/10.5194/isprs-archives-XLIV-M-2-2020-95-2020

method is still acceptable due to its easy access and cheap expense. Surveyors can easily use this method to create maps to observe the interior of a cave, or any other enclosed area.

ACKNOWLEDGEMENTS

We acknowledge Dr. Stacey D. Lyle, Adrian Martinez, Inner Space Caverns, Jennifer and Stephen Johnson for their support, mentorship, and hospitality.

REFERENCES

Gurram, P., Lach, S., Saber, E., Rhody, H., & Kerekes, J. (2007, October). 3d scene reconstruction through a fusion of passive video and lidar imagery. In *36th Applied Imagery Pattern Recognition Workshop (aipr 2007)* (pp. 133-138). IEEE.

Yoo, H.W., Druml, N., Brunner, D. et al. Elektrotech. Inftech. (2018) 135: 408. https://doi.org/10.1007/s00502-018-0635-2

Pix4DMapper Development Team, 2019. Pix4DMapper Software, Version 4.5. Pix4DMapper Development Team. https://www.pix4d.com/product/pix4dmapper-photogrammetry-software (7 November 2019)

CloudCompare Development Team, 2019. CloudCompare Software, Version 2.10.2. CloudCompare Development Team. https://www.cloudcompare.net/ (7 November 2019)

Takahashi, K., 2019. Pcx. https://doi.org/10.5281/zenodo.3560333

Rapidlasso Development Team, 2019. LASTools Software. rapidlasso GmbH. https://rapidlasso.com/lastools/ (7 November 2019)

Unity Technologies, 2019. Unity Software, Version 2019.2.13. Unity Technologies. https://unity.com/ (7 November 2019)

This contribution has been peer-reviewed.
https://doi.org/10.5194/isprs-archives-XLIV-M-2-2020-95-2020

IDENTIFYING EPIPHYTES IN DRONES PHOTOS WITH A CONDITIONAL GENERATIVE ADVERSARIAL NETWORK (C-GAN)

A. Shashank [1], V.V. Sajithvariyar [1,*], V Sowmya [1], K.P. Soman [1], R. Sivanpillai, [2], G. K. Brown [3]

[1] Center for Computational Engineering and Networking, Amrita School of Engineering, Amrita Vishwa Vidyapeetham, Coimbatore, TN 641 112, India - shashankanivilla@gmail.com, (vv_sajithvariyar, v_sowmya)@cb.amrita.edu, kp_soman@amrita.edu
[2] Wyoming GIS Center, University of Wyoming, Laramie, WY 82072, USA (ORCID: 0000-0003-3547-9464) – sivan@uwyo.edu
[3] Department of Botany, University of Wyoming, Laramie, WY 82072, USA – GKBrown@uwyo.edu

KEY WORDS: UAV, CNN, PIX2PIX, UNET, Segmentation, Image translation.

ABSTRACT:

Unmanned Aerial Vehicle (UAV) missions often collect large volumes of imagery data. However, not all images will have useful information, or be of sufficient quality. Manually sorting these images and selecting useful data are both time consuming and prone to interpreter bias. Deep neural network algorithms are capable of processing large image datasets and can be trained to identify specific targets. Generative Adversarial Networks (GANs) consist of two competing networks, *Generator* and *Discriminator* that can analyze, capture, and copy the variations within a given dataset. In this study, we selected a variant of GAN called Conditional-GAN that incorporates an additional label parameter, for identifying epiphytes in photos acquired by a UAV in forests within Costa Rica. We trained the network with 70%, 80%, and 90% of 119 photos containing the target epiphyte, *Werauhia kupperiana* (Bromeliaceae) and validated the algorithm's performance using a validation data that were not used for training. The accuracy of the output was measured using structural similarity index measure (SSIM) index and histogram correlation (HC) coefficient. Results obtained in this study indicated that the output images generated by C-GAN were similar (average SSIM = 0.89 – 0.91 and average HC 0.97 – 0.99) to the analyst annotated images. However, C-GAN had difficulty to identify when the target plant was away from the camera, was not well lit, or covered by other plants. Results obtained in this study demonstrate the potential of C-GAN to reduce the time spent by botanists to identity epiphytes in images acquired by UAVs.

1. INTRODUCTION

Unmanned Aerial Vehicles (UAVs) are used for collecting imagery data on epiphytes that often grow in locations that are difficult to reach from the ground. However, UAV or drone missions can generate large volumes of data. Manually sorting and identifying epiphytes (target plant) in these images can be time consuming and tedious due to the presence of other non-target plants and trees, and background (e.g., horizon). When more images have to be manually processed by one or more analysts it will lead to interpretation errors due to consistency issues (Acevedo, 2020). Under these circumstances, machine learning algorithms such as deep neural networks (DNNs) can be used for identifying epiphytes in these images. Neural networks have been used for identifying plants and trees (Sun et al. 2017) and diseases in plant leaves (Singh and Mishra 2017). DNNs consist of a family of algorithms and new improved ones are introduced periodically. These along with the advances in computing hardware have reduced the time required to process large volumes of image data.

Generative Adversarial Networks (GAN) is one of the advanced DNN algorithms that is capable of recognizing objects or features in the input data and generate output images containing the required information (Goodfellow et al., 2014). GAN were used successfully to diverse applications such as human pose and face generation (Ma and Zhou, 2019), object detection in remotely sensed imagery (Luo and Ding, 2019), text to image synthesis (Qiao et al., 2019), and anomaly detection in time series data (Sun et al., 2019).

Conditional GAN (C-GAN) is a supervised learning algorithm that uses semantic segmentation technique and generates mask by mapping the target on a pixel-by-pixel learning process and excludes the background information (Isola et al., 2017). In this study, we tested C-GAN for identifying epiphytes in 119 images acquired by drones. In the first phase of this project, we trained the C-GAN using a subset of these images. Pixels corresponding to the target plant were highlighted by masking. Using the validation process, we evaluated the network's ability to correctly identify the target plants in the images that were not used for training.

2. MATERIALS AND METHODS

2.1 Aerial images of epiphytes

Aerial images used in this study were acquired in Costa Rica with Phantom DJ(™) and Phantom Sparc(™). More than 2000 aerial images of epiphytes and other vegetation were acquired by a team of researchers (Sivanpillai et al., 2019). These images were sorted into four groups. The first group had images that contained the target epiphyte along with other vegetation. They did not contain any other epiphytes. The second group had images that contained another epiphyte that resembled the target plant along with other vegetation. The third group consisted of images that had several types of epiphytes

* Corresponding author

and in the last group of images only non-epiphytes were present. Images (n = 119) from the first group were used in the first phase of this study. These images were acquired at different times of the day under a range of illumination conditions (Figure 1). Conditions of the target plant also varied in terms of their size, color, and proximity to the camera.

Figure 1. Photos of *W. kupperiana* acquired in bright light (top left), and low light (top right). The target plant was farther (bottom left), and close (bottom right) to the camera.

Numerous images of epiphytes acquired in different lighting conditions and distances are required to train the neural network in order to correctly identify the target plant. However, the number of samples used in this study is relatively less in comparison to the size used by studies that have used this and similar CNN algorithms.

2.2 Image Annotation

Masks corresponding to epiphytes, the target plant, were annotated using LabelMe software (Russell et al. 2008). All parts of the target plant were included in this mask and the rest of the pixels corresponding to other plants and background were assigned to the background. However dried parts (leaves for example) and other debris that obscured parts of the target plant were not included in the mask (Figure 2). Pixels that are highlighted in red color will be recognized as epiphytes by CGAN during the training and validation phases.

Figure 2. Aerial image (left) of *W. kupperiana* and the annotated image (right) with the mask (red) containing pixels of the target plant. Dried parts of the plants were not included in the masks.

2.3 Conditional Generative Adversarial Network (C-GAN)

Conditional GAN (C-GAN) depicts the mapping from an observed (input) image ***x***, and a random noise vector ***z*** to an output ***y***, which can be represented as G: {x, z} = y. In this study, ***x*** will be the drone-acquired image of the epiphytes, and ***y*** will be the annotated image.

GANs are generative models that learn to map from random noise vector z to output image y, G: z → y (Goodfellow et al., 2014). Conditional GANs, on the other hand, learn to map from the observed image x and random noise vector z, to y, G: {x, z} → y.

C-GAN consists of *Generator* and *Discriminator* networks that compete with each other. In the first epoch the *Discriminator* will have a copy of original image and corresponding analyst annotated image. The *Generator* will have analyst annotated image(x) and adds random noise vector (z) to create y. During each iteration, the *Generator* will run a batch of images and generate fake output y and later *Discriminator* will distinguish the previous output from *Generator* and give the feedback. This feedback will be given as back propagation with discriminator loss and the *Generator* will continue this with next batch of image to a point where the *Discriminator* fails to identify the image as fake. This output image is then passed back to the *Discriminator* for identifying the annotated image (correct) from the one (fake) generated by the *Generator*. For every correct identification, *Discriminator* receives a probability score of 1. Results from the first epoch are passed as feedback to the *Generator*.

In the next iteration, Generator learns to create new and better output images using the feedback and passes them to the *Discriminator* for identifying the correct image. In this iteration, the *Discriminator* might incorrectly identify few outputs rather than the correct annotated images. The score will be less than 1 and the feedback is passed back to the *Generator*. As the number of iterations increase, the output images created by the *Generator* will be close to the corresponding annotated image. *Discriminator* will misidentify more and more output images instead of the correct images and its score at the end of iteration will be lower and lower. This process is repeated until the *Discriminator* is unable to identify the correct image. In other word the output image generated by the *Generator* is close to the annotated image.

C-GAN's objective is to balance the competition between the Discriminator and Generator networks. On one hand, the Generator aims to reduce the chances of the output images as fake by the Discriminator. On the other hand, the Discriminator wants to increase the confidence of correctly identifying the output image as fake and this is called the adversarial. This can be expressed as Equation (1):

$$L_{cgan}(G, D) = E_{x,y}[logD(x, y)] + E_{x,z}[\log(1 - D(x, z))] \quad (1)$$

Where, **G** refers to the Generator network that tries to minimize the objective against adversarial D.
D refers to the Discriminative network.
E (x, y) is the expectation value of input x and y output from the generator.
D (x, y) is the discriminator estimation of the probability of x, y to predict real data
E (x, y) is the expectation of mapping the data x with random noise.

https://doi.org/10.5194/isprs-archives-XLIV-M-2-2020-99-2020 | © Authors 2020. CC BY 4.0 License.

D (G (x, y)) is the discriminator estimation of the probability that a fake instance is real.

In C-GAN, G tries to minimize its objective against an adversarial D which tries to maximize its objective. The loss function formulated for the above objective function is a L1 distance which is calculated from Equation (2).

$$L_{L1}(G) = E_{x,y,z}[\|y - G(x,z)\|_1] \quad (2)$$

Lower L1 distance value results in less blurring of the output image. Further details on the formulation of the objective and loss functions are described by Isola et al., 2017.

The final objective function for the C-GAN with L1 distance as loss was calculated using Equation (3).

$$G^* = \arg\min_G \max_D L_{cGAN}(G, D) + \lambda L_{L1}(G) \quad (3)$$

G* refers to a minimum with respect to G, and maximum with respect to D in Equation (3).

C-GAN learns the features for distinguishing epiphytes from the background vegetation in each input image. During the training process the *Generator* network will learn the best weights / features for its network filters to represent the epiphyte data. At the same time, the *Discriminator* will learn the weights to discriminate the images created by *Generator* from the annotated image.

2.3.1 Generator Architecture

The generator network uses an Encoder-Decoder structure as the base for its network architecture. The Encoder network maps the input data to bottle neck junctions where the data is represented in a latent space. The Decoder network maps the compressed latent space data back to transformed out/reconstructed output. The basic structure of an Encoder-Decoder network is represented in Figure 3.

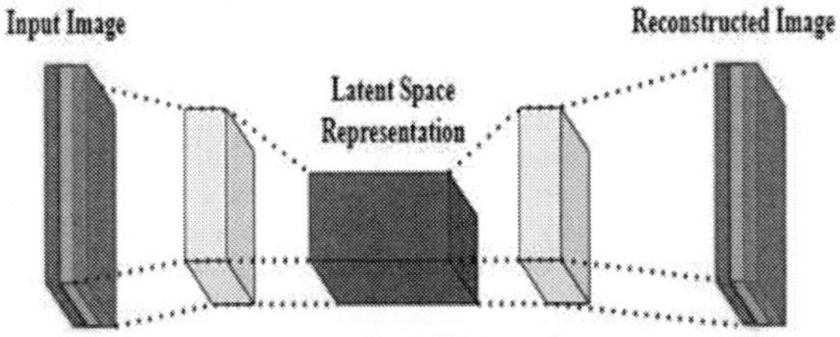

Figure 3: Schematic of the Encoder-Decoder structure which maps an input image to latent space representation and back to a reconstructed image. The red block will be the input image and the CNN filters will derive more fine tunes features in hidden layers (GREEN and BLUE) and form a latent space representation.

The above architecture passes the input image through a series of layers and down sample the data once it reaches the latent space and then it reverses the process. In the study, the input is a RGB image consisting of epiphytes mixed with other vegetation. Later this input image will be transformed with annotated images with red and black colored pixels. In this transformation the structure of the images remains the same but the pixel colors are transformed. This can be considered as a colorization problem where the edge information is very important. The mapping of a low-level information back to high level information must make sure that the relevant features are learned during the training phase. This study used the UNET architecture developed by Ronneberger et al. (2015), with skip connections between the intermediate layers. The skips connections will help the network to learn specific features mentioned earlier for colorization problems. This modification will adapt the Generator network for the translation of input RGB images to annotated images. Figure 4 depicts the structure of an UNET architecture with skip connections.

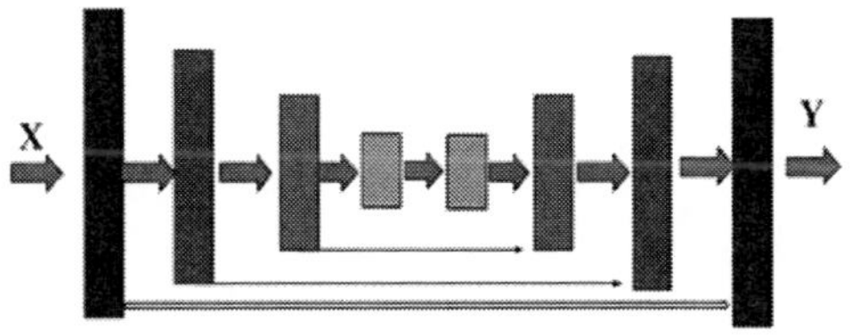

Figure 4: UNET architecture with skip connections that transforms the input RGB image to output images. The image X will be passed through series of hidden layers with many CNN filters, which helps to derive the most import minimum information to represent the original image. The Skip connections will help the CNN filters to learn the weights faster from intermediate layers.

The UNET is built upon a fully convolutional network concept which enables the network to map the input image to annotated image. The major neural network components responsible for carrying the coarse contextual information to higher layers through skip connects are depicted in Figure 5.

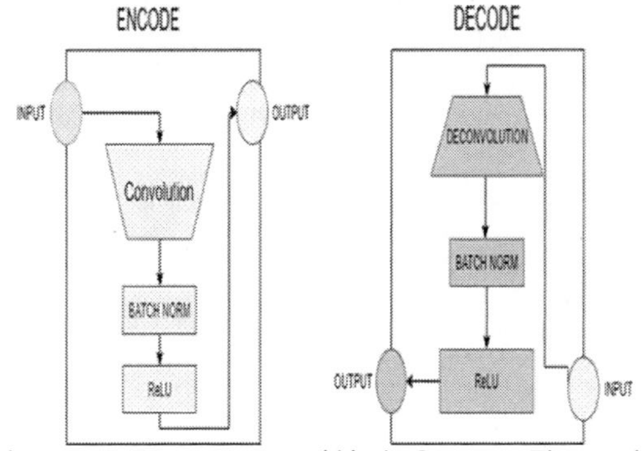

Figure 5: UNET components within the Generator. The encoder block consists of convolutional layers to derive the features, and later weights will be normalized in batch normalization. The ReLU, the fundamental activation function, is used to bring the non-linearity in feature extraction. The decoder block will reverse the process in encoder block and receives the original data back from the encoded features.

2.3.2 Discriminator Architecture

PatchGAN architecture described by Isola et al., (2017) was adapted for classifying the output images as real or fake. The patch GAN network, unlike classifying the entire image as real or fake the images will be divided to small patches and classify it as real or fake. The ideal patch size for best classification found to be 70x70 from experiment work done by Isola et al., (2017), and the same size was used in this study. The schematic representation of the Discriminator PatchGAN is shown in Figure 6.

https://doi.org/10.5194/isprs-archives-XLIV-M-2-2020-99-2020 | © Authors 2020. CC BY 4.0 License.

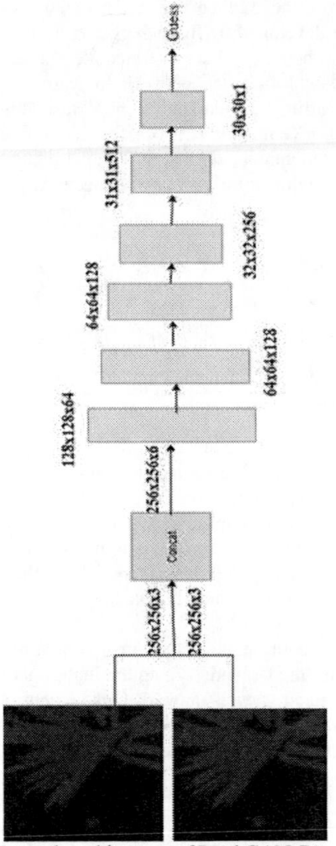

Figure 6. Network architecture of PatchGAN *Discriminator*. One of the two images is the analyst annotated image, while the second image is created by the *Generator*. These images are passed through a series of filters of different kernel size and numbers for feature extraction and comparison.

2.3.3 Training C-GAN

The C-GAN architecture was implemented in python using TensorFlow libraries. The models were trained in a Linux server with i7 Processor, 16 GB of RAM and NVIDIA Tesla K40 GPU. During the training phase, one network will train at a time.

The *Generator* will create a batch of output images and will stop. Next, the *Discriminator* will distinguish the real images from fake ones and will update the *Generator* weight according to the loss returned. This process will continue until the *Generator* and *Discriminator* reaches a point when they share the loss i.e., 0.5 or 50%. Further, there are many hyper parameters that can be tuned to improve the performance of the network. This study considered the number of training samples and iterations for training or epochs. The rest of the next configuration was set as defined by Isola et al., (2017).

Training the *Discriminator* starts with observing a batch of output images created by the *Generator* to find similarity with the analyst annotated image and returns a loss value based on the performance of identifying the fake and real images. The *Discriminator* will also update the weights through backpropagation based on the loss values. *Discriminator* connects with two loss functions which are Discriminator Loss and Generator Loss. During the training, it focuses on its own loss and ignores the Generator loss because it memorizes the weights of input and then refers to its knowledge during the generator training.

2.3.4 Learning Process

The learning curve for the neural network will be linear to the amount of data used for training. The model will be trained for different with the data split into train and validation sets to understand the learning curve. This study explored 3 different splits for training and validation sets: 70% -30%, 80% - 20%, and 90% -10%.

The *Generator* and *Discriminator* networks will be trained for different number of iterations called epochs. The generator will progress with better outputs with the epochs and this will get saturated after certain iterations. The effect of the number of epochs can be analysed from the *Generator*, *Discriminator*, and Objective loss function plots.

2.4 Evaluation Metrics

Output images generated for the validation dataset were evaluated with two metrics: Structural Similarity Index Measure (SSIM), and Histogram Correlation. SSIM measures the similarity between two images (Wang et al., 2004). In this study the output created for the images that were not included in the training were compared to the corresponding annotated images. This index ranges between -1 and 1. A value of 1 indicates that both images are identical and -1 indicates an opposite relationship. Histogram Correlation compares multi-color images using the distribution of their colour intensity values (Swain, Ballard, 1991). Distribution of the red colour pixels corresponding to epiphytes in the output images should be close to the distribution in the output images generated by C-GAN for the validation images. Histogram Correlation computes the Pearson correlation between these two images and the values range between 0 and 1. A value close to 1 indicates that both images are highly correlated.

3. RESULTS AND DISCUSSION

Annotating these 119 images required 35 days because of the complex shape, pattern, and proximity of non-target plants. This is the time consuming and tedious task which required careful examination of each image. Errors introduced in this stage can adversely affect the algorithm's ability to correctly identify the target plants in each image.

Data split (Train% - Test%)	Epoch	SSIM	Histogram correlation
Case 1 (70-30)	300	0.89	0.97
	500	0.90	0.99
Case 2 (80-20)	300	0.90	0.97
	500	0.91	0.97
Case 3 (90-10)	300	0.90	0.97
	500	0.92	0.97

Table 1: The Structural Similarity Index Measure (SSIM) and Histogram correlation values obtained by comparing the annotated and output images generated by C-GAN when 70%, 80%, and 90% of the input images were used for training the algorithm under 300 and 500 epochs.

The average SSIM values obtained by comparing the output images generated by C-GAN with the manually annotated images are presented in Table 1. The average SSIM values ranged between 0.89 and 0.92 and were close to the maximum value of 1. There was a slight increase in the SSIM values when the percent of photos used for training, i.e., data split increased from 70% to 90%. Also, the average SSIM values increased by a fraction between the 300 and 500 epochs.

The average Histogram correlation values were also close to the maximum value of 1 (Table 1). Histogram correlation value increased between 300 and 500 epochs when 70% of the input images were used for training.

The high SSIM and Histogram correlation values that C-GAN algorithm can be trained to identify the target epiphyte plant in the drone-acquired images. Training the network for 500 epochs did not result in major improvement when 70%, 80% and 90% of the input images were used for training.

Sample output images generated by C-GAN were trained with 90% of the input images are shown in Figure 7. These output images were very close to the analyst annotated images.

Figure 7. Output images generated by C-GAN (middle column) and the analyst annotated images (right column). These outputs were generated after the C-GAN algorithm was trained with 90% of the drone acquired images.

Some of the output images generated by C-GAN did not capture all the details annotated by the analyst (Figure 8). The input images were of poor quality due to low lit conditions, location of the target plant within the frame, or occluded by other vegetation. These conditions could have caused the model to either miss some parts of the target plant or incorrectly identify non-epiphytes as the target plant.

Additional pre-processing might be required to improve the quality of these images prior to identifying epiphytes with C-GAN.

Figure 8. Three examples of when the output generated by the algorithm (middle column) failed to correctly identify the target plants as highlighted in the annotated image (right column). This could be due to the target plant was occluded by other vegetation (A), the target plant was not lit sufficiently (B) or was further away from the camera (C).

4. CONCLUSIONS AND RECOMMENDATIONS

C-GAN can be used for identifying epiphytes in these aerial images. This algorithm can help botanists to automatically classify the field data and reduce the time needed to manually identify the target in each photo.

This study also identified few limitations associated with this technique and future work must improve C-GAN's ability to distinguish epiphytes in poor quality images. Incorporating pre-processing techniques such as cropping, scaling, and morphological operations to image before training the algorithm. Future studies can explore the use of other encoder-decoder combinations to improve the overall performance of the network.

This study had access to about 120 photos of the target plants. Most C-GAN studies use 1000s of photos for training the algorithm. Augmentation techniques will help to generate more varieties of training samples and thereby increase the sample size.

Annotating the images was both tedious and time consuming. Future studies must explore automated annotation methods to reduce the time needed to generate the annotated images and increase the consistency.

ACKNOWLEDGEMENTS

UAV images used in this study were acquired through a seed grant from the UW College of Arts & Sciences. We thank Dr. Mario Blanco, University of Costa Rica, and Dr. Carlos de la Rosa, Director, La Selva Research Station for logistic support.

Authors thank Dr. Poornachandran, Centre for Cyber Security Systems, Amritapuri Campus, Kollam, Kerala for supporting the computational platforms. We thank Mr. Sandip Patil and Mr. Harisankar for their help with the CGAN algorithm.

REFERENCES

Acevedo, M. A., Beaudrot, L., Melendez-Ackerman, E., Tremblay, R., 2020: Local extinction risk under climate change in a neotropical asymmetrically dispersed epiphyte. *Journal of Ecology*, 108:1553–1564. doi.org/10.1111/1365-2745.13361.

Goodfellow, I., Pouget-Abadie, J., Mirza, M., Xu, B., Warde-Farley, D., Ozair, S., Courville, A., Bengio, Y., 2014: Generative Adversarial nets. *Proceedings of the Neural Information Processing Systems 2014*, doi.org/papers.nips.cc/paper/5423-generative-adversarial-nets.

Isola, P., Zhu, JY., Zhou, T., Efros, A.A., 2017: Image-to-image translation with conditional adversarial networks. *IEEE Conference on Computer Vision and Pattern Recognition (CVPR)*, 5967-5976, doi.org/10.1109/CVPR.2017.632.

Lou, Z., Ding, S., 2019: Object detection in remote sensing images based on GAN. *International Conference on Artificial Intelligence and Computer Science (AICS 2019)*, 499-503. doi.org/10.1145/3349341.3349458.

Ma, J., Zhou, F., 2019: Multi-poses face frontalization based on pose weighted GAN. *IIEEE 3rd Information Technology, Networking, Electronic and Automation Control Conference* (ITNEC), 1271-1276. doi.org/10.1109/ITNEC.2019.8729088.

Qaio, T., Zhang, J., Xu, D., Tao, D., 2019: MirrorGAN: Learning text-to-image generation by redescription. *IEEE/CVF Conference on Computer Vision and Pattern Recognition (CVPR)*, 1505-1514, doi.org/10.1109/CVPR.2019.00160.

Ronneberger, O., Fischer, P. Brox, T., 2015: U-net: Convolutional networks for biomedical image segmentation. *International Conference on Medical image computing and computer-assisted intervention (MICCAI-2015)*, 234-241. doi.org/10.1007/978-3-319-24574-4_28.

Russell, B.C., Torralbe, A., Murphy, K.P., Freeman, W.T., 2008. LabelMe: A database and web-based tool for image annotation. *International Journal of Computer vision*, 77, 157-173. doi.org/10.1007/s11263-007-0090-8.

Singh, V., Misra, A.K., Detection of plant leaf diseases using image segmentation and soft computing techniques. *Information Processing in Agriculture*, 4(1), 41-49. doi.org/10.1016/j.inpa.2016.10.005.

Sivanpillai, R., Brown, G.K., Gellis, B.S., 2019. Flying UAVs in constrained environments: Best practices for flying within complex forest canopies. In *Applications of small unmanned aircraft systems: best practices and case studies*. Ed. Sharma, J.B., CRC Press, New York, NY. pp:269-282.

Sun, Y., Liu, Y., Wang., G., Zhang, H., 2017: Deep Learning for Plant Identification in Natural Environment. *Computational Intelligence and Neurosciences*. Article ID 7361042. doi.org/10.1155/2017/7361042.

Sun, Y., Yu, W., Chen, Y., Kadam, A., 2019: Time series anomaly detection based on GAN, *Sixth International Conference on Social Networks Analysis, Management and Security (SNAMS)*, 375-382. doi.org/10.1109/SNAMS.2019.8931714.

Swain M.J., Ballard, D.H., 1991: Color indexing. *International journal of computer vision,* 7(1), 11-22. doi.org/10.1007/BF00130487.

Wang, Z., Bovik, A. C., Sheikh, H. R., Simoncelli, E. P., 2004: Image quality assessment: from error visibility to structural similarity. *IEEE transactions on image processing*, *13*(4), 600-612.doi.org/10.1109/tip.2003.819861.

The International Archives of the Photogrammetry, Remote Sensing and Spatial Information Sciences, Volume XLIV-M-2-2020, 2020
ASPRS 2020 Annual Conference Virtual Technical Program, 22–26 June 2020

SUNBATHED ASPEN GROW TO SHOW HOW SUNLIGHT INFLUENCES ASPEN LEAF CHANGES IN THE AUTUMN SEASON

S. K. Weidler [1], R. Sivanpillai [2] *

[1] University of Wyoming, Department of Geography, Laramie, Wyoming 82071, USA, ORCID: 0000-0001-7092-7782 - sweidler@uwyo.edu
[2] Wyoming GIS Center, University of Wyoming, Laramie, WY 82071, USA, ORCID: 0000-0003-3547-9464 - sivan@uwyo.edu

KEY WORDS: Phenology, Fall, Deciduous trees, Laramie, Active learning, TBL

ABSTRACT:

Every autumn, leaves of deciduous trees change from green to other colors and eventually drop to the ground. The rate of color change is influenced by a several factors including the amount of sunlight and temperature. As part of an inquiry-based learning activity, University of Wyoming students have been recording leaf color change (% change) and its drop date (%) in Aspen trees (*Populus tremuloides*) growing in Laramie (WY) using NEON's (National Ecology Observation Network) Phenology data form. In this study, the data recorded from 2015 through 2018 were analyzed to identify trends in the rate of color change in dry and normal years. Trees that were in an area with a high amount of shade were observed to change leaf color and drop their leaves faster than those in areas that received more sun. This pattern was consistent even in years that experienced winter-like conditions in September. Findings from this multi-year study indicate that future environmental modeling projects must factor in the amount of sunlight received by aspen trees in the growing season into account.

1. INTRODUCTION

Aspen (*Populus tremuloides*) trees are found throughout North America (Grant and Mitton, 2010) and they grow in a wide range of climates, temperature, and precipitation regimes. Aspen stands are important habitat for many animals, insects, and other plants (Ripple et al., 2001; Larsen, Ripple, 2003). They serve as an important food source for deer, elk, and beavers. Beavers (*Castor canadensis*) use aspen for building dams and lodges (Beier, Barrett, 1987) which in turn increases species diversity near these locations. In certain geographic areas, domesticated animals (cattle and sheep) also rely on aspen as a source of nutrition (Kaufmann et al., 2013). Given their ecological and environmental impacts, it is important to track the conditions and the environments in which they grow.

Aspen follow an annual phenology cycle starting from bud burst (spring), to maximum leaf area (summer), to senescence (fall) and eventually drop their leaves in late fall (Meier et al., 2015). Timing of these events are influenced by several biotic and abiotic factors (Currit, St Clair, 2010; Hassan and Rahman, 2013). Variations in the timing of the events such as late bud burst, or early senescence will impact other species (Dreiss, Volin, 2013). Aspen plays a role in many aspects of the climate and environment, including the carbon and water cycles (Richardson et al., 2013). Hence, monitoring the phenology of aspen is important to understanding its role in the ecological system in which it is growing (Beaubien, Freeland, 2000). Knowledge of the factors affecting the length of the growing season will serve to predict relationships between the species found in the ecosystem and their impact on the environment.

The Earth's climate, along with its many ecosystems, is rapidly changing. These changes can be hard to track, even with a continual effort towards creating accurate models of the future climate. However, it is necessary to monitor the changes seen in different ecosystem and climate variables such as the levels of carbon dioxide in the atmosphere. Though the task is difficult, it is important to create the most accurate models possible. The ability to predict these changes is very useful because it provides us with the opportunity to better prepare to respond to these changes. These models require an accurate and complete understanding of all the factors that go into them, which includes the role of trees such as aspen. Tracking the variations in the phenology of individual trees could make environment and climate models more accurate than they would have been otherwise (Richardson et al., 2013). For example, phenology affects the growth season of an aspen tree by impacting its senescence, which in turn affects how much water, carbon, and other nutrients are being used by the tree. Understanding how various factors in the tree's environment impact its growing season—such as the amount of light the tree is receiving—can help to better predict how the tree will interact with its environment, and from there how it will impact global climate as a whole.

Figure 1. Some of the aspen trees growing in the University of Wyoming campus in Laramie, WY (USA). Some trees received sunlight most of the day (left) while others were shaded by trees and buildings (right).

* Corresponding author

This paper describes how the amount of sunlight received by aspen trees growing in the University of Wyoming campus affects their phenology and senescence. Students enrolled in an applied remote sensing class monitored deciduous trees during the fall or autumn season (September through early November) for rates of leaf color change and drop. As part of a class exercise, students selected aspen, cottonwood, oak, and maple trees and tracked their phenology on a weekly basis. These trees were growing in open areas or under the shade of buildings and even other trees growing near them. This caused variation in the amount of sunlight received by the trees selected for observation. Some trees received sunlight only in the morning while others in the evening. In few cases, they received very little sunlight throughout the day because they were surrounded by other trees. These observations have been collected since 2015.

The primary objective of this study was to analyze the relationship between the amount of sunlight received by aspen trees and the rates of leaf color change and drop. Every year at least one student team observed a pair of aspen trees – one growing in mostly open space and another growing in a shaded area. During the study period (2015 – 2018), the weather conditions were varying. The secondary objective of the study was to determine if weather conditions (precipitation and temperature), along with varying amounts of sunlight received, influenced the rate of leaf color change and drop in these trees. Findings from this study will provide valuable insights for refining ecological and environmental models.

2. METHODS

2.1 Study area

Phenology of aspen trees growing on the University of Wyoming campus located in Laramie, Wyoming (elevation: 2184 meters or 7165 feet) were analyzed in this study. Laramie is in south-eastern Wyoming, to the east of the Snowy Range of the Medicine Bow Mountains. Climate conditions are warm and dry during the summer months, and cold, snowy, and windy during winter months. The growing season in the area (a period of consistent non-freezing temperatures) starts at the end of May and ends in mid-September, with freezing temperatures and snow often beginning in October. Large temperature and precipitation fluctuations are common in autumn. Maximum daily temperatures in autumn can range from 4.7°C to 24.2°C.

2.2 Data

Students in the applied remote sensing course (BOT 5130/4130) recorded leaf color change and drop in NEON's (National Ecology Observation Network) Citizen Science Bud Burst phenology forms (Figure 2). Since observations were made in the autumn season, only data on 'Leaves changing color' and 'Leaves dropping' were recorded. Leaf color changes in the selected aspen trees were recorded as i) No leaves have changed color, ii) Early – only a few leaves have changed color (< 5%), iii) Middle – many leaves have changed color, or iv) Late – most leaves have changed color (> 95%). Every year, students recorded their observations until all the leaves had dropped from the trees.

In addition to leaf color and drop, students were also required to record the geographic coordinates of the trees they monitored along with a short description of the locations. From these observations, we selected four sets of datasheets containing phenology records of eight aspen trees: four growing in a relatively sunny (open) area and the other four in shady areas.

Temperature and precipitation data recorded at the Laramie airport for 2015 – 2018 were downloaded from the Weather Underground (wunderground.com/history) for relating the observed phenological changes in aspen trees.

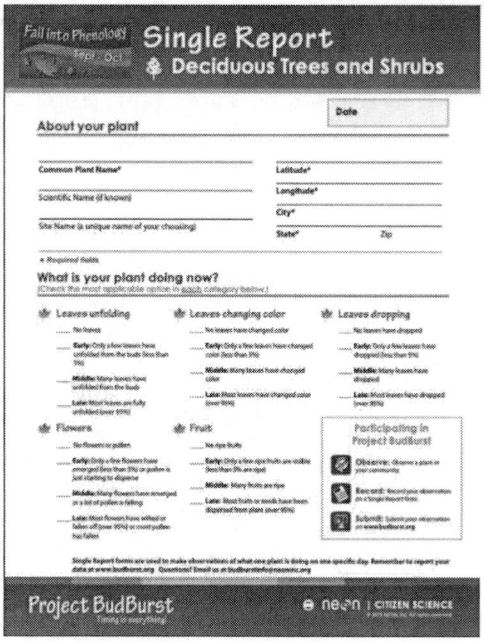

Figure 2. Phenology data form used by the students enrolled in the applied remote sensing course at University of Wyoming for tracking leaf color change and drop of deciduous trees. This form was developed by National Ecology Observation Network (NEON) as part of the Project BudBurst project.

Sudden drop in temperature, especially below 0° C, in early fall could influence the phenology of the deciduous trees. Similarly, early season precipitation such as snowfall can stress the leaves and branches of a tree. Almost all trees growing on-campus are irrigated at least once a week along with the grass lawns, which reduced the effect of days with low amounts of precipitation.

2.3 Interpretation

Observation dates were converted to Day of the Year (DOY) for respective years. Next, data on 'Leaves changing color' and 'Leaves dropping' were plotted against DOY using percentage of color change or leaf drop. One plot was made for each year to visualize differences between pairs of aspen trees that grew in different amounts of sunlight.

Average minimum temperatures were plotted against each month starting from September through November. Total precipitation was also plotted for each month. Historic averages for each of the three months were also included to provide additional context for interpreting the monthly precipitation values.

Plots of leaf color change and drop were visually compared to average minimum temperature and total precipitation plots to identify patterns and potential relationships between weather and phenological patterns. These were compared because sudden drop in temperature or early season precipitation in the form of snow could influence the phenology of aspen trees.

3. RESULTS

3.1 Precipitation and Temperature Trend

Plot of average minimum temperatures recorded for 2015 – 2018 (Figure 3) revealed distinct temperature patterns for the 3 autumn months.

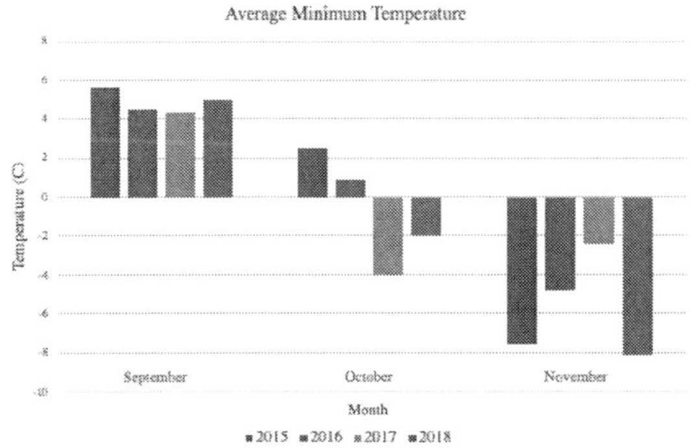

Figure 3. Monthly average minimum temperature recorded for September, October, and November 2015 – 2018.

Temperature values gradually declined every month in 2015, 2016, and 2018 with little variations from day to day (not shown in figure). In 2017 there was a rapid decline in temperature in October and then a slight increase in November.

September 2016 had relatively more days when the minimum temperature dropped below the freezing point compared to the rest of the years analyzed in this study (Table 1).

Month	2015	2016	2017	2018
September	0	7	4	2
October	10	14	26	23
November	28	26	20	29

Table 1. Number of days in each month in 2015 – 2018 fall seasons when the minimum temperature fell below 0°C (32°F).

In September 2015, the minimum temperature never dropped below the freezing point. October 2015 had fewer days of below freezing temperatures than October of other years (Table 1).

Monthly precipitation totals recorded for each of the months for 2015 – 2018 are shown in Figure 4. The historical average precipitation received in these months was also included for comparison.

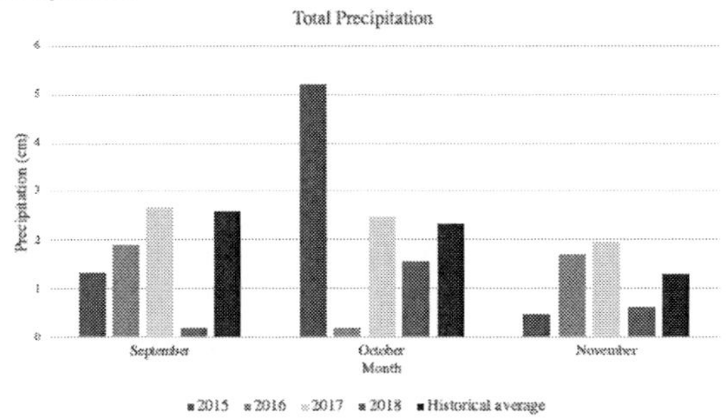

Figure 4. Total monthly precipitation recorded for September, October, and November 2015 – 2018, along with the historical averages for each month.

Laramie received average precipitation in September, October, and November 2017. September and November 2018 had very little precipitation in comparison to the average values. In October 2018 there were several large precipitation events in the beginning of the month. Compared to the historic average, there was twice the amount of precipitation in October 2015. In October 2016 there was relatively low precipitation compared to the historic average for that month.

3.2 Leaf Color Change and Drop

In autumn 2017, the color of the aspen leaves in the tree growing in the shaded area (blue bar in Figure 5) changed approximately a week earlier than the tree exposed to more sunlight (yellow bar). Additionally, the aspen tree growing in the shaded area (blue line) dropped its leaves a week earlier than the tree exposed to more sunlight (yellow line).

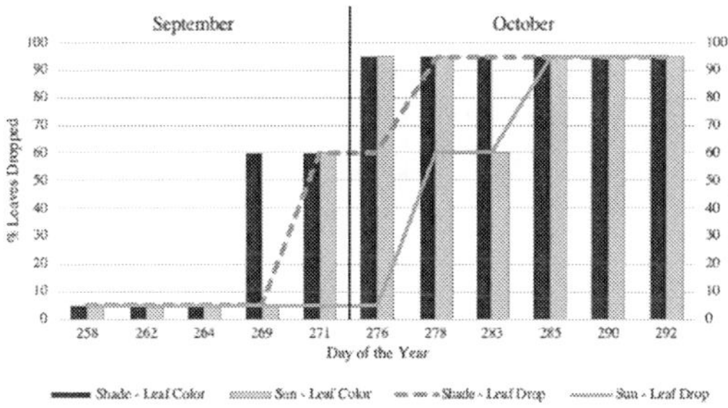

Figure 5. Temporal changes in leaf color change (bars) and leaf drop (lines) recorded in fall 2017 in the aspen trees growing in shaded (blue) and sunny (yellow) areas.

On DOY 283, this team had recorded 65% leaf color change for the sun exposed aspen tree. However, on DOY 276 and 278 this team had already recorded 95% leaf color change for the same tree. Hence the entry on DOY 283 must be transcribing error.

A similar pattern of leaf color change was noticed for a different pair of aspen trees in 2016 except for one unusual observation on DOY 272 (Figure 6). Leaves of the aspen trees that grew in shade (blue bars) changed their color more quickly in comparison to the trees that received more sunlight (yellow bars).

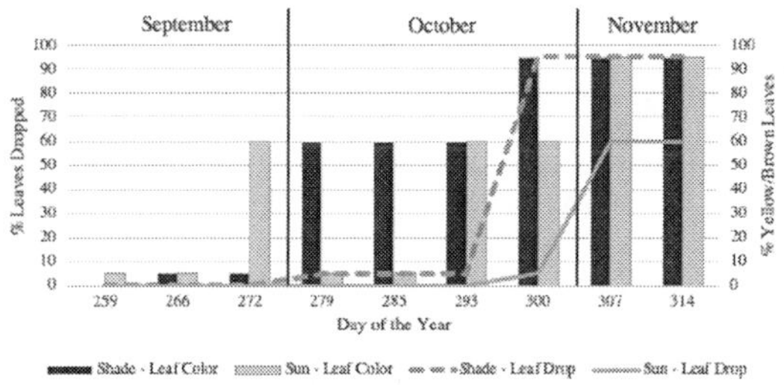

Figure 6. Temporal changes in leaf color change (bars) and leaf drop (lines) recorded in fall 2016 in the aspen trees growing in shaded (blue) and sunny (yellow) areas.

On DOY 272, the team noted that 65% of the leaves in the aspen tree receiving more sunlight changed color, but on subsequent observations (DOY 279 and 285) recorded that there was 5% color change. Hence the observation recorded for the sunny tree on DOY 272 was excluded from interpretation. Leaves of the shaded tree (blue line) started to drop approximately a week earlier than the leaves of the tree receiving more sunlight (yellow lines in Figure 6). In comparison to 2017 (Figure 5), the leaves dropped later in both trees in 2016.

The pattern of leaf color change and drop in 2018 was similar to the patterns recorded in the two previous years, except both processes occurred earlier in comparison to the years before (Figure 7). The shaded tree began changing color a week earlier and was losing its leaves at a more rapid rate than the sunny tree by the end of data collection. However, the total number of observations were fewer in the 2018 season because data were not collected at the same frequency as in the 2016 and 2017 seasons.

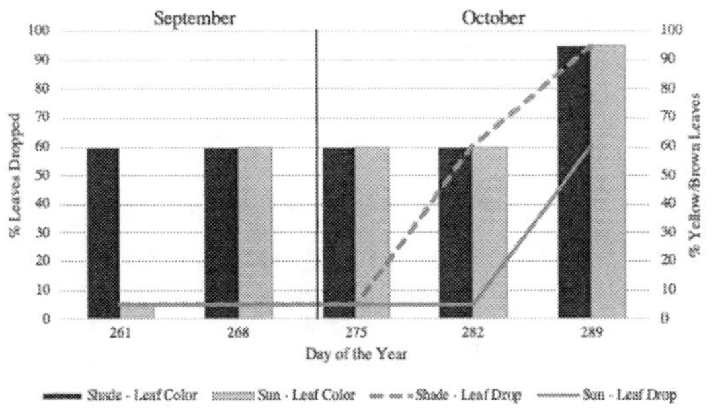

Figure 7. Temporal changes in leaf color change (lines) and leaf drop (bars) recorded in fall 2018 in the aspen trees growing in shaded (blue) and sunny (yellow) areas.

Observations recorded in 2015 (Figure 8) for another pair of aspen trees showed a delayed onset of color change (DOY 288) for both trees. Similarly, leaf drop started on DOY 302 for both trees. Coincidently, the minimum temperature never dropped below freezing point in September 2015.

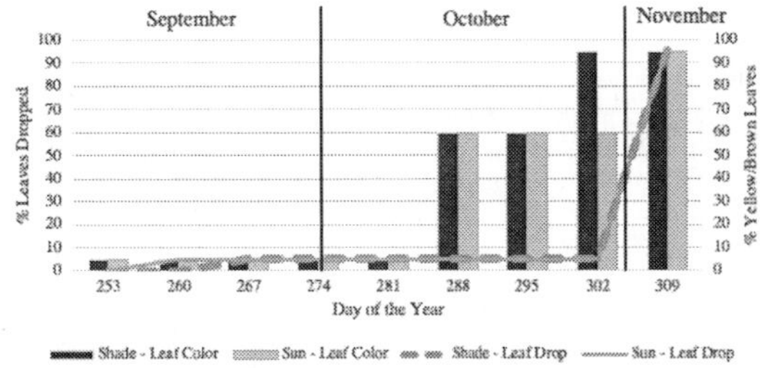

Figure 8. Temporal changes in leaf color change (bars) and leaf drop (lines) recorded in fall 2015 in the aspen trees growing in shaded (blue) and sunny (yellow) areas.

October 2015 had fewer days with the minimum temperature dropping below the freezing point than other years. Despite the warmer temperature recorded in the first two months, the shaded tree changed its color one week before the sunny tree. However, both trees dropped their leaves at approximately same time.

4. DISCUSSION

4.1 Multi-year phenology trends

Aspen trees that received relatively less sunlight in spring and summer changed their leaf color and dropped their leaves earlier in the year than those trees that received more sunlight. This pattern was consistent in the data collected by students enrolled in this course over four years (2015 – 2018), despite the differences in data collection dates and frequency between groups.

This trend was also consistent under varying weather conditions each year. Despite having different precipitation and day-to-day temperature patterns every year, trees in the shade always lost their leaves and changed color more rapidly than their sunnier counterpart.

4.2 Data collection and interpretation

Interpreting data collected in multiple years by different groups of students posed some challenges. Not all groups collected data on the same day of the week and some groups even collected data more than once a week. Few of the groups randomly collected data on their trees rather than following once-in-a-week pattern of other groups. In some cases, the frequency of data collection decreased as fall senescence set in. If more time had passed between two consecutive data collections, it is possible to miss the rate of leaf color change or drop. Within the same year, some teams had recorded 11 observations, and a few had recorded only three. Data recorded on random days for few weeks would have missed some phenological stages. In this study, we excluded data collected for fewer weeks. If all teams had recorded sufficient data, phenology data from more pairs of trees could have been analyzed.

For future years, all teams must be required to collect data for both trees at same interval—once a week for example, and preferably on the same day of the week. Once senescence sets in, teams must collect data twice a week to capture detailed phenological changes. Instructors must require students to submit the completed forms every week to check for any inconsistencies. This would allow students to correct their data through additional observations. Consistent data collection between groups and years will enable direct comparisons of changes throughout senescence.

The phenology data form developed by NEON has only four categories of leaf color changes and drop. Limiting the responses to four categories introduced some ambiguity. For example, each group could have interpreted the categories 2 (few), 3 (many), and 4 (most) slightly differently. In some groups, students took turns collecting data which introduced errors. For example, in one week a member would record a value of Middle (60% change) for leaf color change, while next week another member would record a value of Early (5% change). To address these inconsistencies, it is important to have consistent definition for each category in order to minimize variations between groups collecting data in the same and different years.

Only very few groups recorded additional information such as the amount of sunlight received in their data forms. Since shady trees receive different amounts of sunlight, their leaf color and leaf drop phenology differ from trees growing in sunnier locations. The groups that recorded the amount of sunlight did so in order to identify light level as the factor of comparison between trees.

NEON forms do not collect this information because their primary purpose is to record the phenology of individual trees. In future years, student teams must record additional details about the shady trees. For example, if a tree receives some sunlight for a short amount of time in the morning or evening it must be recorded as part of the summary observations.

This contribution has been peer-reviewed.
https://doi.org/10.5194/isprs-archives-XLIV-M-2-2020-105-2020 | © Authors 2020. CC BY 4.0 License.

We recommend that additional information on temperature and precipitation be recorded in the weekly datasheets. This information will be helpful for interpreting changes in leaf color and drop, especially when abrupt changes occur. For example, when the temperature suddenly drops or if the study area receives early fall season snow, this information will be valuable to interpret any rapid changes in leaf color or drop. This information can be used to account for the effect of weather conditions on the same trees in multiple years.

In this study, phenology data for four different pairs of aspen trees were analyzed for as many years. In the future, phenological changes in one pair of aspen trees must be tracked over multiple years. This will provide more insights about how the same pair of trees changed their leaf color in different years.

4.3 Tree location data

Some groups recorded the geographic coordinates more precisely than others. Also, many teams used their smart devices (phones) to record the geographic coordinates. The precision of these locational data entered in the data forms was not sufficient to locate all the trees, especially if they were growing near other aspen trees of similar size or age. However, many groups had included brief description regarding the location in addition to the coordinates. To reduce ambiguity in future studies, teams must include an identification tag before taking photos of the tree. This can allow future teams to precisely locate the trees for multi-year studies.

4.4 Other Patterns

In fall 2015, the shady aspen tree monitored by one of the groups was surrounded by other aspen trees, i.e., grew in a cluster. This group observed that this tree changed leaf color quicker and dropped its leaves earlier than a tree that grew by itself away from other trees. This points out that additional factors, such as proximity to other aspen trees, must be considered in the future. It would be useful to identify whether growing in a group is something that impacts aspen every year regardless of temperature and precipitation. We recommend that students must be encouraged to choose this factor for data collection in future semesters.

In the future, the phenology data for trees growing in varying sunlight levels could be expanded to other species of trees growing on the University of Wyoming campus. There are several cottonwood trees (*Populus sect. Aigeiros)* of different age and size, that are growing in different light conditions which could result in a rich set of phenology data. It would be important to examine whether other species follow a similar or different trend of earlier senescence in trees which have grown in areas with less exposure to sunlight.

5. CONCLUSIONS

Phenology data of pairs of aspen trees recorded from 2015 to 2018 indicate that the amount of sunlight received by aspen trees influenced color changes of leaves and rate of leaf drop. This pattern was consistent across all four years though there were variations in data collection dates and frequency within and across these years.

Warmer fall seasons (2015) as measured by fewer number of days where the minimum temperature dropped below the freezing point delayed the changes by few weeks.

Budburst forms developed by NEON are able to track major phenological changes but adding a few more categories for leaf color and drop would increase the precision of the recorded data.

Including additional information about the trees such as the amount of sunlight they receive during the day, their proximity to other trees, and photos with recognizable markers will enable more detailed interpretation of phenology data. This will also enable future teams to precisely locate the same pairs of trees for multi-year observations.

ACKNOWLEDGEMENTS

First author acknowledges the WyomingView internship opportunity that made this study possible. Additionally, thanks to the Wyoming Geographic Information Science Center (WyGISC) and the Department of Geography at the University of Wyoming for providing the funding to travel and present this study at the ASPRS GeoWeek 2020 conference.

This material is based upon work supported by the U.S. Geological Survey under Grant/Cooperative Agreement No. G18AP00077 to the second author. The views and conclusions contained in this document are those of the authors and should not be interpreted as representing the opinions or policies of the U.S. Geological Survey. Mention of trade names or commercial products does not constitute their endorsement by the U.S. Geological Survey. This manuscript is submitted for publication with the understanding that the United States Government is authorized to reproduce and distribute reprints for Governmental purposes.

REFERENCES

Beaubien, E.G., Freeland, H.J., 2000. Spring phenology trends in Alberta, Canada: links to ocean temperature. *International Journal of Biometeorology,* 44:53–59. DOI:

Beier, P., Barrett, R.H., 1987. Beaver Habitat Use and Impact in Truckee River Basin, California. *The Journal of Wildlife Management*, 51(4): 794-799.

Currit, N., St Clair., S.B., 2010. Assessing the impact of extreme climatic events on aspen defoliation using MODIS imagery. *Geocarto International.* 25(2): 133-147. DOI: 10.1080/10106040903051975.

Dreiss, L.M., Volin, J.C., 2013. Influence of leaf phenology and site nitrogen on invasive species establishment in temperate deciduous forest understories. *Forest Ecology and Management*, 296: 1-8. DOI: 10.1016/j.foreco.2013.01.031.

Grant, M., Mitton, J., 2010. Case Study: The Glorious, Golden, and Gigantic Quaking Aspen. *Nature Education Knowledge* 3(10):40.

Hassan, Q.K., Rahman, K.M., 2013. Applicability of remote sensing-based surface temperature regimes in determining deciduous phenology over boreal forest. *Journal of Plant Ecology*, 6(1): 84–91. DOI: 10.1093/jpe/rts010.

Kaufmann, J., Bork, E.W., Alexander, M.J., Blenis, P.V., 2013. Habitat selection by cattle in Foothill landscapes following variable harvest of aspen forest, *Forest Ecology and Management*, 306: 15-22. DOI: 10.1016/j.foreco.2013.06.004.

Larsen, E.J., Ripple W.J., 2003. Aspen age structure in the northern Yellowstone ecosystem: USA. *Forest Ecology and Management*, 179: 469-482. DOI: 10.1016/S0378-1127(02)00532-7.

Meier, G.A., Brown, J.F., Evelsizer, R.J., Vogelmann, J.E., 2015. Phenology and climate relationships in aspen (*Populus tremuloides* Michx.) forest and woodland communities of southwestern Colorado, *Ecological Indicators*, 48: 189-197. DOI: 10.1016/j.ecolind.2014.05.033.

Richardson, A.D., Keenan, T.F., Migliavacca, M., Ryu, Y., Sonnentag, O., Toomey, M., 2013. Climate change, phenology, and phenological control of vegetation feedbacks to the climate system. *Agricultural and Forest Meteorology*, 169:156-173. DOI: 10.1016/j.agrformet.2012.09.012.

Ripple, W.J., Larsen, E.J., Renkin, R.A., Smith, D.W., 2001. Trophic cascades among wolves, elk and aspen on Yellowstone National Park's northern range, Biological Conservation, 102(3) 227-234. DOI: 10.1016/S0006-3207(01)00107-0.

BIG IMAGERY AND HIGH PERFORMANCE COMPUTING AS RESOURCES TO UNDERSTAND CHANGING ARCTIC POLYGONAL TUNDRA

C.Witharana [1*], M. A. E. Bhuiyan [1], A. K. Liljedahl [2]

[1] Dept. of Natural Resources and the Environment, University of Connecticut, Storrs, Connecticut, USA - (chandi.witharana, md.bhuiyan)@uconn.edu
[2] Woods Hole Research Center, Falmouth, Massachusetts, USA - aliljedahl@whrc.org

KEY WORDS: Arctic, Permafrost, Ice-wedge polygons, Deep learning, Satellite imagery, Mapping

ABSTRACT:

Permafrost thaw has been observed at several locations across the Arctic tundra in recent decades; however, the pan-Arctic extent and spatiotemporal dynamics of thaw remains poorly explained. Thaw-induced differential ground subsidence and dramatic microtopographic transitions, such as transformation of low-centered ice-wedge polygons (IWPs) into high-centered IWPs can be characterized using very high spatial resolution (VHSR) commercial satellite imagery. Arctic researchers demand for an accurate estimate of the distribution of IWPs and their status across the tundra domain. The entire Arctic has been imaged in 0.5m resolution by commercial satellite sensors; however, mapping efforts are yet limited to small scales and confined to manual or semi-automated methods. Knowledge discovery through artificial intelligence (AI), big imagery, and high performance computing (HPC) resources is just starting to be realized in Arctic science. Large-scale deployment of VHSR imagery resources requires sophisticated computational approaches to automated image interpretation coupled with efficient use of HPC resources. We are in the process of developing an automated Mapping Application for Permafrost Land Environment (MAPLE) by combining big imagery, AI, and HPC resources. The MAPLE uses deep learning (DL) convolutional neural nets (CNNs) algorithms on HPCs to automatically map IWPs from VHSR commercial satellite imagery across large geographic domains. We trained and tasked a DLCNN semantic object instance segmentation algorithm to automatically classify IWPs from VHSR satellite imagery. Overall, our findings demonstrate the robust performances of IWP mapping algorithm in diverse tundra landscapes and lay a firm foundation for its operational-level application in repeated documentation of circumpolar permafrost disturbances.

1. INTRODUCTION

Arctic permafrost - unique landscapes comprising the Earth materials that remains at or below 0°C for at least two consecutive years - occupies approximately 24% of the exposed land surface of the northern hemisphere. Ice-rich permafrost can be identified by atypical surface features called ice-wedge polygons (IWPs), which are underlain by several meter-wide and deep ice-wedges that form a network across the tundra.

Thawing of ice-rich permafrost can be seen from satellites by altered moisture and vegetation that follows the differential ground subsidence when the top of the ice-wedges thaws. The diameter of IWPs typically ranges from 5 m to 30 m and the associated microtopographic features, such as troughs or rims, are in sub-meter to ~1-meter scale (Kanevskiy et al. 2017). Vegetation and geology maps suggest that about two-thirds or more of the Arctic landscape is occupied by polygonal ground (Raynolds et al. 2019) and therefore ice-rich ground, but the exact extent and the prevailing IWP types (i.e. whether the ice wedges experienced melt or not) are largely unknown (Liljedahl et al. 2016). The microtopography associated with IWPs affects the Arctic ecosystem from local to regional scales due to the impacts on the flow and storage of water and therefore, vegetation and carbon. Over recent decades, ice-wedge degradation – transformation of low-centered polygons into high-centered polygons due to ice-wedge degradation has been documented at several locations across the Arctic tundra in the field and through localized remote sensing analyses (Steedman et al., 2017). The shift from one IWP type to the other is documented to occur in less than a decade (Liljedahl et al. 2016) with an unusual warm summer, wildfires, or human activities initiating the onset of ice-wedge degradation. Subsequent feedback processes can either continue, amplify, dampen or even reverse the ice-wedge melt, with such processes remaining active for multiple decades. Degradation of ice wedges is a quasi-cyclic process with degradation often occurring over a shorter time scale than the formation of new permafrost (aggradation), with the latter controlled by accumulation of organic and mineral soil above the ice-wedge (Kanevskiy et al. 2017). Understanding of spatiotemporal dynamics behind the evolution of ice-wedge polygonal tundra demands for objective and detailed maps consolidating the ice wedge extent and their prevailing successional stages.

Despite the alarming signals, yet, the Arctic science community has a limited understanding of the spatiotemporal continuity of the otherwise locally observed changes. The lack of knowledge on the larger geographical extent and successional stage of IWPs introduce uncertainties to regional and pan-Arctic estimates of carbon, water, and energy fluxes. Remote sensing provides transformational opportunities to observe, monitor, and measure the Arctic polygonal landscape at multiple spatial scales and varying temporal windows. IWPs are difficult to detect in any remote sensing imagery with spatial resolution greater than 4 m. Therefore, sub-meter resolution commercial satellite imagery (e.g. DigitalGlobe, Inc.) demonstrate a greater promise in accurate delineation and characterization of ice-wedge polygonal networks. Due to IWPs' varying spectral and morphometric characteristics visual inspection and manual digitization has so far been the most widely adopted and promising method to delineate polygons from high resolution remote sensing imagery. A considerable number of local-scale studies have analyzed ice wedge degradation processes using satellite imagery, and manned-/unmanned aerial imagery/LiDAR data (Muster et al. 2013). Most of the studies

This contribution has been peer-reviewed.
https://doi.org/10.5194/isprs-archives-XLIV-M-2-2020-111-2020

to date have relied on manual image interpretation and/or semi-automated approaches (Skurikhin et al. 2014). Therefore, there is a need and an opportunity for utilization of VHSR imagery in regional scale mapping efforts to spatio-temporally document microtopographic changes due to thawing ice-rich permafrost.

Despite the accuracy, the intense workload of manual digitization constrains the size of a study area. Sophisticated image analysis algorithms are therefore needed to automatically extract ice-wedge polygons and their prevailing stages using hundreds to thousands of commercial satellite imagery. Recently, deep learning (DL) convolutional neural nets (CNNs) has shown great potential for object instance segmentation in detecting and delineating each distinct object in an image of common objects from everyday pictures. The success of DLCNNs in computer visions (CV) applications has received great interest from the remote sensing community. There has been an explosion of studies integrating DCLNN to address remote sensing classification problems spanning from general land use and land cover mapping to targeted feature extraction. Deep learning CNNs excel at object detection, semantic segmentation, and semantic object instance segmentation. A plethora of DLCNN architectures have been proposed, developed, and tested. The influx of new DLCNNs continues to grow. Each has its own merits and disadvantages with respect to the detection and/or classification problem at hand. DL-based approaches are being adapted to finely classify Arctic permafrost features from high-resolution remote sensing imagery (aerial (Zhang et al. 2018), satellite (Bhuiyan et al. 2019), imagery-derived digital elevation models (DEMs), and LiDAR-based DEMs.

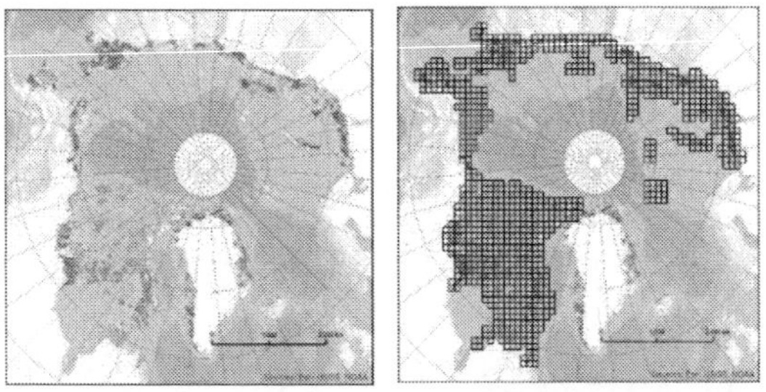

Figure 1. (a) Circumpolar Arctic vegetation map (CAVM) showing the extent of tundra (Raynolds et al. 2019). (b) A 100 km x 100 km grid overlain on CAVM to show the geographic extent to be mapped using commercial satellite imagery, comprising tundra of Alaska, Canada, and Russia. The area under the grid covers approximately 5 million km^2.

The entire Arctic has been imaged at 0.5 m resolution by commercial satellite sensors (DigitalGlobe, Inc.). The image repository at the Polar Geospatial Center (PGC) provides transformational opportunities to observe, monitor, and document permafrost thaw occurring across the Arctic, which is a remote region with extremely sparse field observation network. Alaska, Canada, and Russia collectively harbor approximately 5 million km^2 of tundra (Figure 1). Producing a permafrost thaw map for the pan-Arctic tundra region just once would require an analysis of thousands commercial satellite image scenes. Such large-scale deployment of imagery resources requires sophisticated computational approaches to automated image interpretation coupled with efficient use of high-performance computing (HPC) resources. Besides our ongoing efforts, to the best of our knowledge, no university-led study has so far been developed a fully automated and scalable method, which is capable of accurately detecting and characterizing surface features from sub-meter satellite imagery over large geographical areas (e.g. pan-Arctic) in an operational context.

The overarching goal of our ongoing effort is the production of the first pan-Arctic ice-wedge polygon map using large volume of commercial satellite imagery available at the Polar Geospatial Center and HPC resources. The resulting circumpolar ice-wedge polygon map will advance our understanding of the complex and interlinked processes responsible for the evolution of the pan-Arctic ice-wedge polygon tundra landscape. Here we unravel a new framework that we developed and tested – Mapping Application for Permafrost Land Environment (MAPLE) – to drive imagery-enabled Arctic permafrost science applications. We also present some of the automated ice-wedge polygon mapping results while relating to the DLCNN model interoperability across different tundra vegetation types. As seen on Figure 1, we have so far completed mapping of over 10000 km^2 of North Slope of Alaska. The rest of the Alaska is being processed at the moment and we will expand the mapping operation to rest of tundra regions.

2. METHODS

2.1 Mapping Application for Permafrost Land Environment

A general schematics of our mapping application for permafrost land environment (MAPLE) is shown in Figure 2. While our primary target is large volumes of commercial satellite imagery, manned- and unmanned-aerial imagery can also be integrated into the workflow as needed. Our target is to derive science-ready products from imagery. Such products are not limited to ice-wedge polygons but extend to other permafrost thaw features and microtopograpy such as thaw slumps, lake erosion, trough, and capillaries or even vegetation (shrubs) mapping. MAPLE primarily sits on HPC resources such XSEDE (eXtreme Science and Engineering Discovery Environment) that are available for the U.S. National Science Foundation (NSF) funded research.

Accurate characterization of IWPs from VHSR imagery directly depend on the segmentation (i.e. isolation of targets from the surrounding) and classification (i.e. assigning the correct label to the targets) processes. Semantic object instance segmentation methods are designed to afford target isolation and labeling to thematic classes. Ideally, a mapping application for permafrost land environment should consist of candidate DLCNN models tailored to extract different permafrost features of interest from remote sensing imagery. Among suite of target features, micrtopography, thaw features, capillaries, and plant functional exhibit high priority. Given the diversity of target features and their heterogeneous characteristics coupled with semantic complexities, multiple model architectures better serve the purpose. In MAPLE, one pipeline targets on mapping ice-wedge polygons in which we utilized Mask RCNN algorithm. The pipeline is extensible and tailored to work with remote sensing imagery using high performance computing resources. This allows scalability to larger spatial extents. Mapping workflow and Mask-RCNN architecture is detailed out in Figure 2. Our mapping workflow is modular. First, input image scene is portioned into tiles of 200 pxl 200 pxl. Next, tiles are streamed to the trained model for prediction. Post processing of predictions takes place in stage 3. The predicted categorical raster is vectorized as a shapefile. All the individual shapefiles are post-processed by omitting duplicates along tile borders and merged together to create a single shapefile corresponding to the extent of the input satellite image scene. The workflow is parallelized and optimized to work on multiple GPUs and nodes.

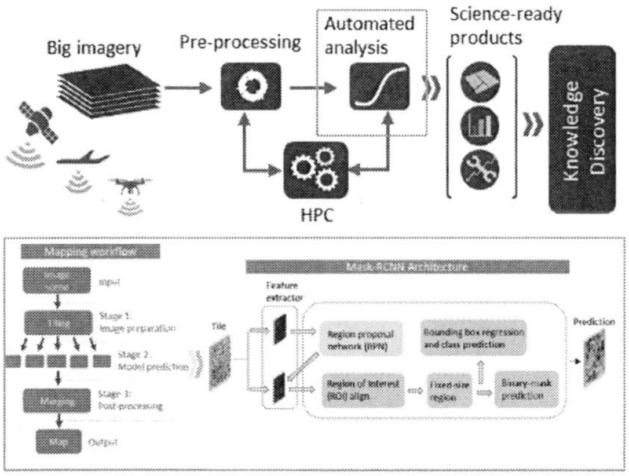

Figure 2. Simplified schematic of the Mapping Application for Permafrost Land Environment (MAPLE) (top). Imagery-enabled workflow embedded with the Mask R-CNN architecture (bottom).

The Mask R-CNN (He et al. 2016) serves as the key DLCNN algorithm in MAPLE. It has widely been acknowledged as a promising algorithm in semantic segmentation tasks across multitude scientific domains. The Mask-RCNN is an extended method for object instance segmentation built on the Faster R-CNN with the aid of a function for predicting masks for distinct objects. The Mask R-CNN generates proposals (i.e., candidate object bounding boxes) after scanning the image, and subsequently, the model predicts the class, bounding box, and binary mask for each region of interest (ROI). The primary segments of the Mask R-CNN include: backbone architecture Residual Learning network (ResNet) for feature extraction, Feature Pyramid Network (FPN) for improving representation of objects at multiple scales, and other modules, such as Region Proposal Network (RPN) for generating (RoI), RoI Classifier for class prediction of each RoI, Bounding Box Regressor (BBR) for refining RoI, and FCN with RoIAlign and bilinear interpolation for predicting pixel-accurate mask.

2.2 Model Training

We practiced transfer learning strategy to re-train the Mask-RCNN network. Annotated data (defining and labelling regions of interest) were created using an online web tool "VGG Image Annotator" from satellite imagery comprising heterogeneous tundra types. We randomly selected 262 cropped subsets (tiles of 200 pxl by 200 pxl) (~15,000 polygons) from different tundra types (tussock, non-tussock, and sedge) considering the spectral, and spatial variability. Each file has 200 x 200 pixels. Dataset are annotated for two classes: Low-centered (LC) polygons (8962 objects) and high-centered (HC) polygons (6038 objects). The annotated tiles were randomly divided into a training dataset, validation dataset, and test dataset based on the 8:1:1 split rule. We trained the Mask R-CNN model with a mini-batch size of two image tiles, 350 steps per epoch, learning rate of 0.001, learning momentum of 0.9, and weight decay of 0.0001. To minimize overfitting, random horizontal flips augmentation was applied to introduce variability in the training data that has acceptable estimation accuracy. During calibration, the weights and biases of each neuron were estimated iteratively by minimizing a mean squared error cost function using a gradient descent algorithm with back propagation. Training was implemented using NVIDIA V100 GPUs on XSEDE supercomputing resources. We trained the Mask R-CNN with 100 epochs. To optimize Mask R-CNN, we examined different losses, such as (a) Smooth-L1 loss, defines box regression on object detection systems, which is less sensitive to outliers, than other regression loss; (b) Mask R-CNN bounding box loss indicates the difference between predicted bounding box correction and true bounding box; (c) Mask R-CNN classifier loss estimates difference of class labels between prediction and ground truth ; (d) mask binary cross-entropy loss measures (probability value between 0 and 1) the performance of a classification model by observing predicted class and actual class; (e) RPN bounding box loss identifies the regression loss of bounding boxes only when there is object and; (f) RPN anchor classifier loss indicates the difference between the predicted(RPN) and actual (closest ground truth box to the anchor box) regression.

2.3 Model Interoperability

We employed a systematic experiment to investigate the model's immunity to landscape heterogeneity across three tundra types (tussock, non-tussock, and sedge). The experiment was based on four summer-time multi-spectral images acquired by the WorldView-2 satellite sensor (Figure 3). Pansharpened multispectral images at 0.5m were provided by the Polar Geospatial Center as orthorectified, atmospherically corrected data products.. Scenes primarily comprised wet-sedge, tussock, and non-tussock tundra according to the Circumpolar Arctic Vegetation map (Raynolds et al. 2019). Candidate scenes cover 1500 km^2 of coastal and upland tundra (tussock, non-tussock, and sedge), from the North Slope, Alaska. We selected training study sites comprising tussock, non-tussock, sedge, and barren tundra dominant landscapes, primarily from coastal tundra region of North Slope Alaska, Canada, and Wrangler Island of Russia. The training sites provide a substantial landscape heterogeneity for model classifying and detection of ice-wedge polygons.

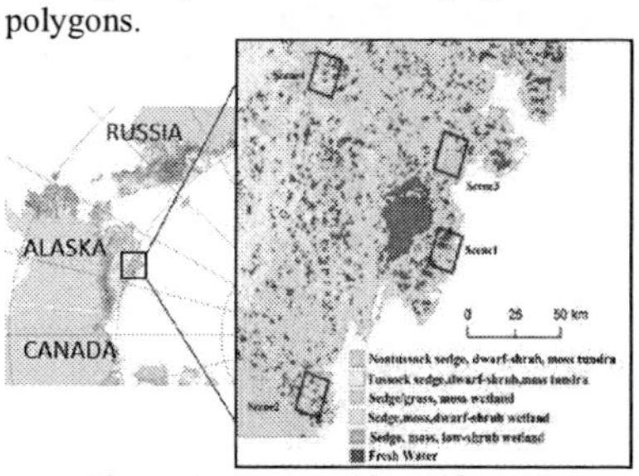

Figure 3. Geographic setting (left) and (b) tundra vegetation map overlain by four candidate satellite image scenes.

2.4 Accuracy Assessment

We tasked several error metrics to assess the DL model performances across tundra vegetation types.

The mean intersection over union (mIoU) between predicted and actual polygon. A mIoU score > 0.5 is considered a "good" prediction which indicates successful delineation.

$$\mathbf{mIoU} = \frac{A_O}{A_U} \qquad \text{............(1)}$$

Absolute mean relative error (AMRE) is the mean of the relative percentage error, calculated by the normalized average

https://doi.org/10.5194/isprs-archives-XLIV-M-2-2020-111-2020 | © Authors 2020. CC BY 4.0 License.

$$AMRE = |\frac{1}{n}\sum_{i=1}^{n}\left(\frac{\hat{y}_i - y_i}{y_i}\right)| \quad \text{..........(2)}$$

Here, umber of predicted polygons $\hat{y}$, the number of actual polygons y_i. For an unbiased model, the AMRE would be 0.

Correctness, which indicates how many of predicted positives were truly positive; Completeness, determines what percentage of actual positives were detected; F1 Score, which determines a balance between Correctness and Completeness into one value [29]. High magnitudes of Correctness suggest that there are less false positives in the classification. In addition, if the model classification always predicts positive magnitudes, Completeness will be 1, which indicates ice-wedge polygon is properly detected by the model. Moreover, an accurate prediction of ice-wedge polygon is represented by F1 score, where score of 1 specifies perfect prediction. An accurate prediction is represented by all metric values closing to 1.

$$\text{Correctness} = \frac{TP}{TP + FP} \quad \text{..........(3)}$$

$$\text{Completeness} = \frac{TP}{TP + FN} \quad \text{..........(4)}$$

$$\text{F1 Score} = \frac{2 * \text{Correctness} * \text{Completeness}}{\text{Completeness} + \text{Completeness}} \quad \text{.....(5)}$$

Here, true positive (TP) is the number of polygons correctly identified, false positive (FP) is the number of polygons identified by model but not true, and false negative (FN) is undetected polygons.

3. RESULTS AND DISCUSSION

We statistically evaluated the performances of the DL model in detecting and classifying IWPs. For the quantitative assessments, from each image scene, we randomly selected 40 subsets to manually delineate polygons as a reference (ground-truth polygons). The mean intersection over union (mIoU) values varied between 0.85 to 0.91 (Table 1), which indicted that predicted polygons that agree with the ground-truth polygons.

Scene	mIoU
S1	0.91
S2	0.87
S3	0.86
S44	0.85

Table 1. Summary Statistics of mean intersection over union

Zoomed-in views of the original imagery, ground truth, and model classification results show that our predicted IWPs closely matched ground-truth IWPs (Figure 4).

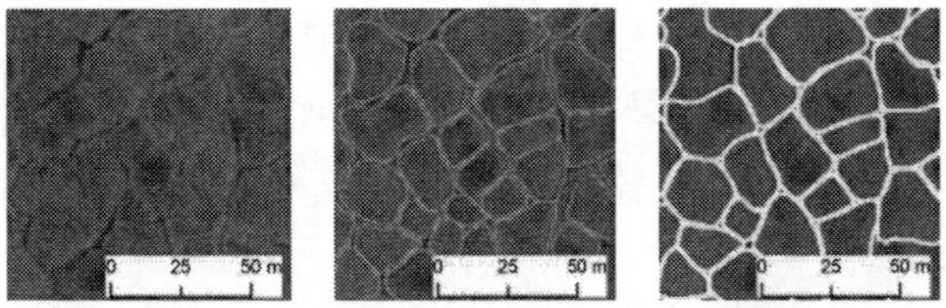

(a) (b) (c)

Figure 4. Zoomed-in views of (a) original imagery, (b) ground truth (manual delineation, blue outline) and (c) model result (yellow outline) for candidate scene 4. Imagery © [2016] DigitalGlobe, Inc.

We used three quantitative error statistics (correctness, completeness, and F1 score) to show the performances of the framework. Candidate scenes 1, 2, 3, and 4 produced high model detection accuracies for the F1 score, ranging from 0.89 to 0.96 (Figure 5, Table 2).

Scene	#of reference polygons		Completeness	F1 Score
S1	582	0.99	89%	0.96
S2	567	1	85%	0.94
S3	579	1	83%	0.92
S4	573	1	81%	0.89

Table 2. Reported scores for correctness, completeness, and F1 measure

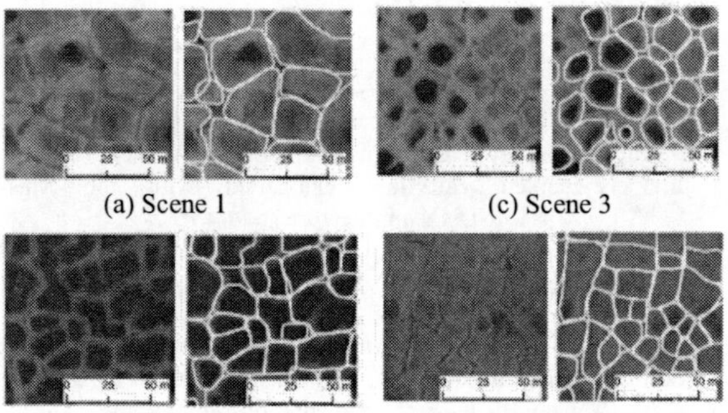

(a) Scene 1 (c) Scene 3

(b) Scene 2 (d) Scene 4

Figure 5. Sample views of original imagery (left) and model classification (right) for candidate scenes. Yellow outlines denote automatically detected IWPs. Imagery © [2010, 2012, 2015, 2016] DigitalGlobe, Inc.

Although all the image scenes are geographically close to each other, but they still have different tundra variations in the microtopography. Predominance of tussock sedge tundra and the high spatial resolution of imagery information provide landscape-scale variation within the original CAVM polygons throughout northern Alaska. Scene 4 (covering tundra tussock sedge) achieves mIoU 0.85 (Table 1) which still have a chance to improve model prediction by increasing more training data from that tundra region. Moreover, lake rich regions, such of Alaska's North Slope demonstrated dominant sedges tundra, which contains more detailed information which will help IWP mapping for that tundra type. Image scenes 2 and 3 represents non-tussock sedge tundra of Alaska's North Slope. Model performances for image scenes 2-3 (F1 score: 0.92-0.94) were consistent, which means training sample were sufficient to predict IWPs for non-tussock sedge tundra regions. In a similar fashion, scenes 1-4 scored high values for completeness (81-89%). In all four cases, the correctness metric scored ~ 1, allowing less freedom for false alarms. Classification accuracies varied from 0.94 to 0.97 for candidate scenes, indicating a robust performance of the new version of the ice-wedge polygon mapping algorithm across different tundra types in northern Alaska. Results showed substantially low systematic errors (AMRE values from 0.17 to 0.23) for all candidate scenes (Figure 6). Overall, both quantitative and qualitative evaluations support the possible interoperability of the IWP mapping algorithm across different tundra assemblages in northern Alaska.

This contribution has been peer-reviewed.
https://doi.org/10.5194/isprs-archives-XLIV-M-2-2020-111-2020

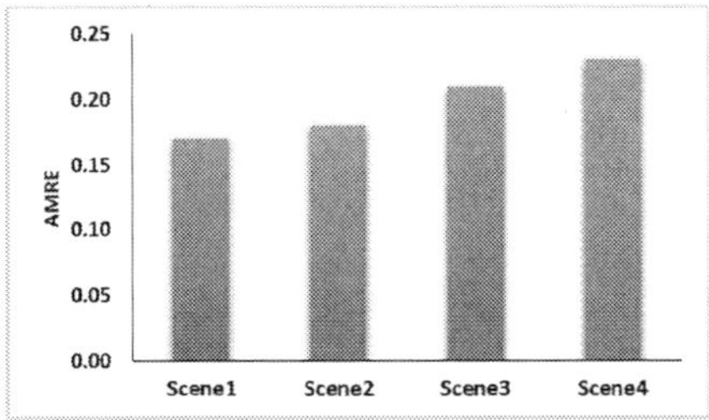

Figure 6. Absolute Mean relative Error (AMRE) for candidate scenes

Considering the geographical extent of Arctic tundra and associated landscape complexities that yet to be exposed, we make no strong claims that our transferability study is comprehensive but an exploratory effort. Arctic tundra landscapes cover spatially isolated ponds, lakes, marshes, river, and stream corridor wetlands, which representing highly heterogeneous features, varying in soil moisture, vegetation composition, elevation, surficial geology, ground ice content, soil thermal regimes and surface hydrology. Fine-scale differences in microtopography, limit the ability to comprehend local scale controls on regional to global scale patterns which, is an important factor in characterizing IWPs in arctic varying tundra areas. As Arctic tundra vegetation are spatially heterogeneous which vary significantly over small spatial scales, including additional different tundra landforms with different vegetation, hydrology, or soil characteristics could further improve the model. Further research is inevitably needed to better understand how trained models behave across tundra types. Such study can incorporate not only vegetation cover but also other factors as well, such as terrain units, soil types, hydro-climatic regimes, and permafrost characteristics. For instance, Arctic tundra types are very sensitive in summer temperature, which can cause major changes to vegetation structure via by pose spectral/textural changes in the acquired imagery. Thus, the seasonality could be an important factor deciding the appearance of ice wedge polygon on the satellite imagery because changes to spectral and textural characteristics can alter the overall semantics of the target. The model can be biased when it is trained only based on the imagery acquired in particular time window. The predictions can be suffered, if the model is given imagery from different time windows, for example early summer image vs late summer image. As much as important to understand model's interoperability across space, it is also imperative to examine how model responds to temporal variations. Operator biasness in hand-annotated data production can also negatively influence model performances. Tasking multiple operators to produce sizeable amount of quality-controlled training datasets can help improving the variability training samples and eventually leveraging the model performances.

4. CONCLUSION

We are in the process of developing a Mapping Application for Permafrost Land Environment (MAPLE) by combining big imagery, AI, and HPC resources. The MAPLE uses DLCNNs algorithms on HPCs to automatically map IWPs from VHSR commercial satellite imagery across large geographic domains. We trained and tasked a DLCNN semantic object instance segmentation algorithm to automatically classify IWPs from VHSR satellite imagery. We explored the DLCNN model interoperability across different tundra types and image scene complexities in order to understand the opportunities and challenges prior to any future circumpolar IWP mapping applications. The DL model exhibited promising performances with high detection and classification accuracies. Also reported low systematic error for all image scenes and indicated significant improvement in model predictions across the heterogeneous tundra regions. Consideration of contextual information (e.g. edges, vegetation, shape, area, and the consistency of feature distributions) increased the reliability of the model classification and helped generalizing the DL model across tundra vegetation types. Complex topography play a vital role in controlling the spatial variation in image scenes. Effort to further refine model prediction accuracies could include a) increasing the variability of training samples with additional annotated IWPs from a larger set of tundra vegetation types, and b) explore more sophisticated image pre-processing steps such as differing data fusion approaches.

ACKNOWLEDGEMENTS

This research was support by the U.S. National Science Foundation grant #: 1722572 and 1720875. Supercomputing resources were provided by the eXtreme Science and Engineering Discovery Environment . Authors would like to thank Polar Geospatial Center, University of Minnesota for imagery support.

REFERENCES

Bhuiyan, M.A.E., Witharana, C. and Liljedahl, A.K.,. 2019. Big Imagery as a Resource to Understand Patterns, Dynamics, and Vulnerability of Arctic Polygonal Tundra. AGUFM, pp.C13E-1374.

He, K.; Zhang, X.; Ren, S.2016. Deep residual learning for image recognition. In Proceedings of the IEEE conference on computer vision and pattern recognition, Las Vegas, America, 27-30 Jun.

Kanevskiy, M.; Shur, Y.; Jorgenson, T.; Brown, D.R.N.; Moskalenko, N.G.; Brown, J.; Walker, D.A.; Raynolds, M.K.; Buchhorn, M.2017. Degradation and stabilization of ice wedges: Implications for assessing risk of thermokarst in northern Alaska. Geomorphology, 297, 20–42.

Liljedahl, A.K.; Boike, J.; Daanen, R.P. ; Fedorov, A.N. ; Frost, G.V. ; Grosse, G. ; Hinzman, L.D. ; Iijma, Y.; Jorgenson, J.C. ; Matveyeva, N. ; Necsoiu, M. 2016. Pan-Arctic ice-wedge degradation in warming permafrost and its influence on tundra hydrology. Nat. Geosci.,, 9, 312.

Muster, S.; Heim, B.; Abnizova, A. ; Boike, J. 2013. Water body distributions across scales: a remote sensing based comparison of three arctic tundra wetlands. Remote Sens., 5, 1498–1523.

Raynolds, M.K.; Walker, D.A.; Balser, A.; Bay, C.; Campbell, M.; Cherosov, M.M.; Daniëls, F.J.; Eidesen, P.B.; Ermokhina, K.A.; Frost, G.V.; Jedrzejek, B. A. 2019. raster version of the Circumpolar Arctic Vegetation Map (CAVM). Remote Sensing of Environment, 232,111297.2019.

Skurikhin, A. N.; Wilson, C. J.; Liljedahl, A.; Rowland, J. C. 2014. Recursive active contours for hierarchical segmentation of wetlands in high-resolution satellite imagery of arctic landscapes. Southwest Symposium on Image Analysis and Interpretation, 137–140.

https://doi.org/10.5194/isprs-archives-XLIV-M-2-2020-111-2020

Steedman, A. E.; Lantz, T. C.; Kokelj, S. V.2017. Spatio-temporal variation in high-centre polygons and ice-wedge melt ponds, Tuktoyaktuk coastlands, Northwest Territories. Permafr. Periglac. Process., 28, 66–78.

Zhang, W.; Witharana, C.; Liljedahl, A.K.; Kanevskiy, M. 2018. Deep Convolutional Neural Networks for Automated Characterization of Arctic Ice-Wedge Polygons in Very High Spatial Resolution Aerial Imagery. Remote Sens, 10, 1487.

CNN-BASED PLACE RECOGNITION TECHNIQUE FOR LIDAR SLAM

Y. Yang *, S. Song, C. Toth

Dept. of Civil, Environmental and Geodetic Engineering, The Ohio State University, 470 Hitchcock Hall, 2070 Neil Avenue
Columbus, OH 43210, USA - (yang.2695, song.1634, toth.2)@osu.edu

KEY WORDS: Lidar, Global descriptor, Deep learning, Place recognition

ABSTRACT:

Place recognition or loop closure is a technique to recognize landmarks and/or scenes visited by a mobile sensing platform previously in an area. The technique is a key function for robustly practicing Simultaneous Localization and Mapping (SLAM) in any environment, including the global positioning system (GPS) denied environment by enabling to perform the global optimization to compensate the drift of dead-reckoning navigation systems. Place recognition in 3D point clouds is a challenging task which is traditionally handled with the aid of other sensors, such as camera and GPS. Unfortunately, visual place recognition techniques may be impacted by changes in illumination and texture, and GPS may perform poorly in urban areas. To mitigate this problem, state-of-art Convolutional Neural Networks (CNNs)-based 3D descriptors may be directly applied to 3D point clouds. In this work, we investigated the performance of different classification strategies utilizing a cutting-edge CNN-based 3D global descriptor (PointNetVLAD) for place recognition task on the Oxford RobotCar dataset[1].

1. INTRODUCTION

One important aspect of SLAM algorithms is that the localization errors keep accumulating as the number of measurements keeps increasing, due to the errors in measurements caused by the noise of sensors (Dhiman et al., 2015). To handle this problem, SLAM algorithms rely on place recognition (PR), or loop closure detection (LCD) techniques, wherein the algorithms are able to recognize previously visited places and then use them as additional constraints for increasing the precision of localization estimation and solving the global localization problem. Therefore, a robust PR scheme could enhance the robustness and performance of SLAM algorithms. For the Lidar-SLAM, PR is still a challenging task and very few of the state-of the art algorithms has solved the loop closure problem (Singandhupe et al., 2019). Many methods have been proposed for this task, and a traditional solution is sensor integration with other sensors, such as camera (Olson, Edwin, 2009a) (Wu et al., 2016) or GPS (Emter, Thomas, 2012) (Emter et al., 2018). However, these techniques face challenges, such as vision based methods suffering from illumination changes, season-to-season based appearance changes and viewpoints differences, and poor GPS performance in urban areas.

Since Lidar data is invariant to lighting and appearance changes, the geometric methods for PR with 3D Lidar data, such as line feature-based scan matching, key point matching and 3D local feature-based strategies are widely investigated (Olson, Edwin, 2009b) (Bosse et al., 2013) (Dubé et al., 2017). Unfortunately, extracting and matching these features could be difficult in certain environments. To that end, CNN-based solutions have recently been proposed as effective learning tools to generate features from general environments. Due to the different ways to learn and extract descriptors, these solutions can be classified in two categories: semantic (local) level feature-based (Dubé et al., 2018) and frame (global) level feature-based (Angelina, Hee Lee, 2018) (Yin H et al., 2018) (Yin H et al., 2019) (Yin P et al., 2018a) (Yin P et al., 2018b). The major limitation for extracting semantic features is the assumption that there are enough static objects which have been adequately learned by the pretrained CNN model. However, this assumption may not always be satisfied in real-world practice. On the other hand, with the global descriptor, the PR task is handled as a similarity modeling problem in which Nearest Neighbor (NN) method is commonly used for classification. Additionally, one interesting task in the real-world PR practice is classification under the restriction that we may only observe a single example of each possible scenario before making a prediction about a test instance. This problem is known as one-shot learning (Koch et al., 2015), and the Siamese neural networks have been demonstrated as an effective solution for one-shot learning in imagery application (Yin W et al., 2015) and low dimensional 3D semantic segment descriptors classification (Cramariuc et al., 2018).

To efficiently generate reliable PR candidates by improving the performance of classification network, in this study we investigated a one-shot learning classification method, the CNN-based Siamese network with high dimensional global descriptors on 3D Lidar data (Figure 1). In the experiment, we compared the effectiveness of classification between our CNN-based classifier, a commonly used nearest neighbor (NN) method and random forests (RF) which is a typical nonlinear classic machine learning classifier. The details of proposed method are discussed in the remainder of this paper, structured as follows. Section 2 reviews the proposed method, including network for global feature descriptor extraction and CNN classifier model. The experiments, including training, testing and performance comparison are presented in Section 3. Finally, the conclusions are summarized in Section 4.

* Corresponding author
[1] https://robotcar-dataset.robots.ox.ac.uk/

2. METHOD

2.1 Global Descriptor

Compared to its image counterpart, applying a CNN model to 3D points is more challenging due to the fact points in a point cloud are generally unordered. Some works handled this challenge by projecting 3D point clouds into 2D image plane (Su et al., 2015) (Yin P et al., 2018b) or transforming point clouds into 3D volumetric representations (Qi et al., 2016) (Yin P et al., 2018a). The downside of these networks is that they cannot handle well the large-scale outdoor PR problems. Additionally, for these networks, Lidar data need to be preprocessed to provide proper input which is computationally expensive. To directly operate on an unordered points subset in a point cloud, (Angelina, Hee Lee, 2018) proposed the PointNetVLAD[2] network which integrates PointNet network and VLAD layer (Figure 2).

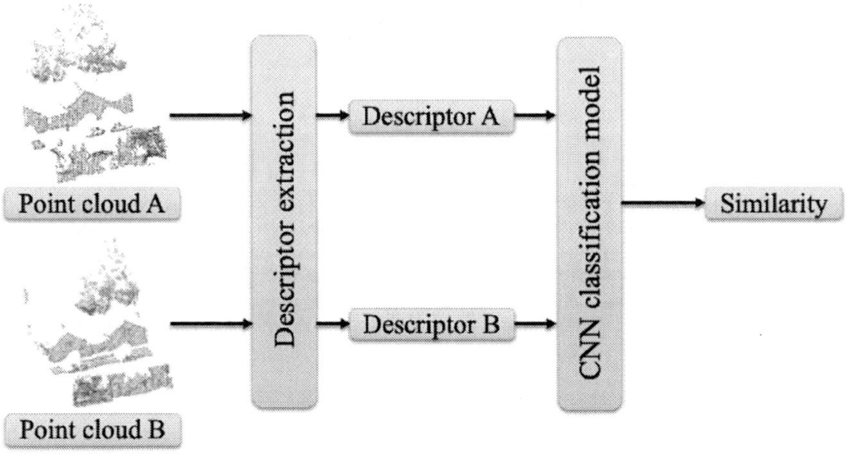

Figure 1. Proposed network for PR task

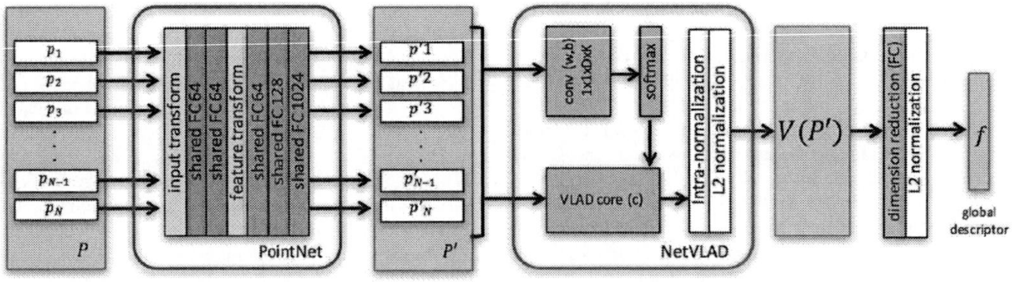

Figure 2. Network architecture of PointNetVLAD (Angelina, Hee Lee, 2018)

The PointNet extracts local feature descriptors for each input point by encoding points into vectors in a higher dimensional space. In the next phase, the NetVLAD layer aggregates local features into the VLAD bag-of-words (BoWs) global feature descriptor vectors. Additionally, since NetVLAD is a symmetric function and PointNet model transforms each point in the point cloud independently, the output global descriptor is invariant to the order of the points. In the training process, PointNetVLAD was trained with the lazy quadruplet loss in which the Euclidean distances between descriptors are used for calculating similarity. During inference (testing), NN method was used for classification. The lazy quadruplet loss is defined as:

$$L = \max_{j}\left(\left[\alpha + \delta_{pos} - \delta_{neg_j}\right]_+\right) + \max_{k}\left(\left[\beta + \delta_{pos} - \delta_{neg_k^*}\right]_+\right) \qquad (1)$$

where α, β = margin
δ = Euclidean distances between global descriptor vectors

In this work, pretrained PointNetVLAD baseline network was used as global feature extractor.

2.2 CNN-based Classifier

As depicted in Fig 1, two input point clouds are firstly given to two branches of the Siamese network which are distinct PointNetVLAD networks and create global descriptors. In the next stage, these two descriptors are combined and processed by a CNN classification model in which the similarity score is calculated as the final output. The structure of CNN classification model is detailed in Figure 3.

[2] https://github.com/mikacuy/pointnetvlad.git

https://doi.org/10.5194/isprs-archives-XLIV-M-2-2020-117-2020

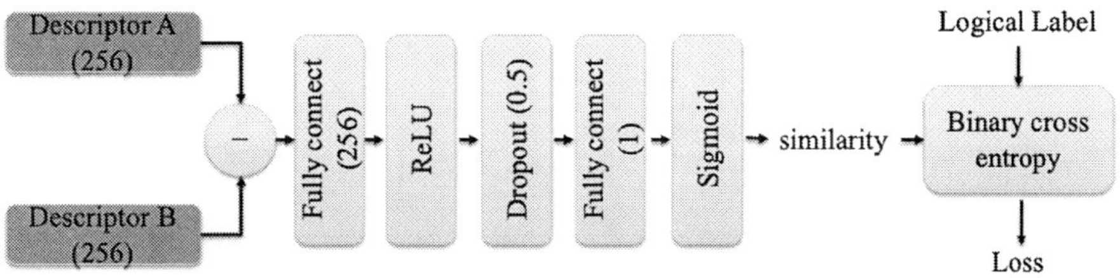

Figure 3. Network architecture of CNN classification model

One advantage of Siamese network is that the feature extraction model and classification model can be trained and used simultaneously or independently. In this work, we trained and used CNN classification model independently for a fair comparison of classification performance with respect to other classifiers. The binary cross entropy loss with stochastic gradient descent (SDG) is applied during training process. The loss function is formed as:

$$L = -t \log y - (1 - t) \log(1 - y) \qquad (1)$$

where t = logic label
y = output of CNN classification model

3. EXPERIMENT

3.1 Training the Model

Since we only investigate the performance of classifiers in this work, the same training and test dataset, as used in original PointNetVLAD research, was applied to guarantee a consistent performance of feature extraction. The dataset was built from the Oxford RobotCar dataset (Maddern et al., 2017) in which 44 sets of full and partial runs were used. Training and testing reference maps are geospatially separated from each run with a proportion of 70% and 30%, respectively. Then submaps were segmented from reference maps following the rules: (1) each submap contains all Lidar points within a 20m trajectory of the vehicle, and (2) the intervals between submaps are 10m and 20m for training and testing datasets, respectively. Finally, 21,711 training submaps and 3,030 testing submaps were segmented out from original dataset. The submaps within 10m intervals in centroid coordinates are seen as structurally similar and labelled as "positive" and those with 50m are dissimilar and "negative".

The training results for the CNN classification model are shown in Figure 4. The loss quickly converged during the first few epochs and remains almost constant in subsequent epochs. Thus, we stopped training at 1000 epochs. The CNN classification model was trained on a Nvidia GeForce Titan Xp GPU.

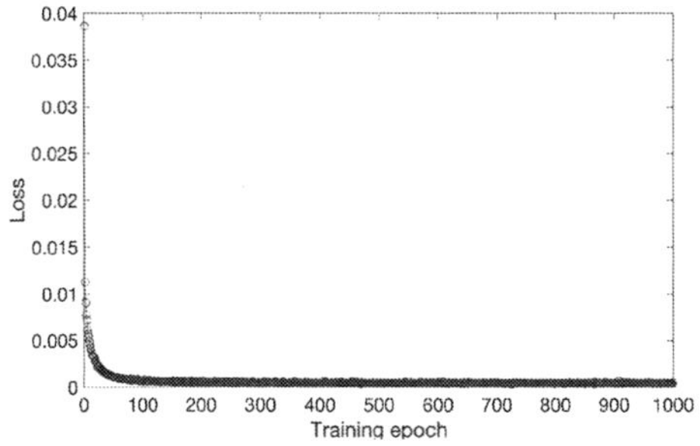

Figure 4. Training results of CNN classification model

3.2 Comparing Different Classification methods

In this section we compare the performance of different classification methods, e.g. CNN-based classifier, NN and RF by using the same set of global feature descriptors extracted from testing datasets. In terms of training the RF, the input is two concatenated descriptors and output is their matching probability. The closest neighbour in NN method is decided based on the Euclidian distance in descriptor vector space. The receiver operating characteristic (ROC) curves of the different classifiers are shown in Figure 5. The best accuracy is achieved by the CNN-based classifier. The numerical results are presented in Table 1. CNN-based classifier outperforms NN and RF in both general accuracy and true positive (recall) rate which are 95.3% and 70.1% respectively. Note that the commonly used NN method performs the worst, achieving a general accuracy of 61.1%, and true positive rate of 40.7%. The classic nonlinear classifier RF, used as a reference, outperforms NN with 88.5% in general accuracy and 57.6% in true positive.

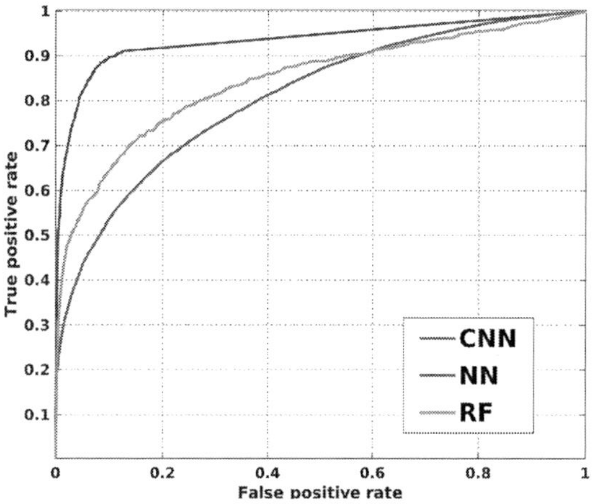

Figure 5. ROC curves for different classifiers with same set of global feature descriptors

Method	General Accuracy	True Positive Rate
NN	61.1%	40.7%
RF	88.5%	57.6%
CNN-based classifier	95.3%	70.1%

Table 1. Matching Accuracy statistics of different classifiers

Examples visualizing the matching results are presented in Figure 6 in which (a)~(c) are true positive matching, (d)~(f) are false positive matching. It can be seen that the proposed method shows robustness to noise, such as objects changing (Fig 6(a)),

This contribution has been peer-reviewed.
https://doi.org/10.5194/isprs-archives-XLIV-M-2-2020-117-2020 | © Authors 2020. CC BY 4.0 License.

viewpoint changing (Fig 6(b)) and both objects and viewpoint changing (Fig 6(c)). On the other hand, Fig 6(d)~(f) reveal the fact that the CNN based classifier fails to distinguish dissimilar submap pairs when two scenarios contain very similar features, such as semblable building structures and trees.

4. CONCLUSION

In this work, we investigated the performance of CNN based classifier with Lidar data for PR task. The testing results show that the proposed model outperforms both NN and RF methods and achieves true positive rate at 70.05%. However, many false matchings occur when scenarios contain very similar features. In the future work, we will try to (1) increase the performance in recall by using geometric or other constrains to reject false matches, and (2) integrate the proposed place recognition method into Lidar SLAM.

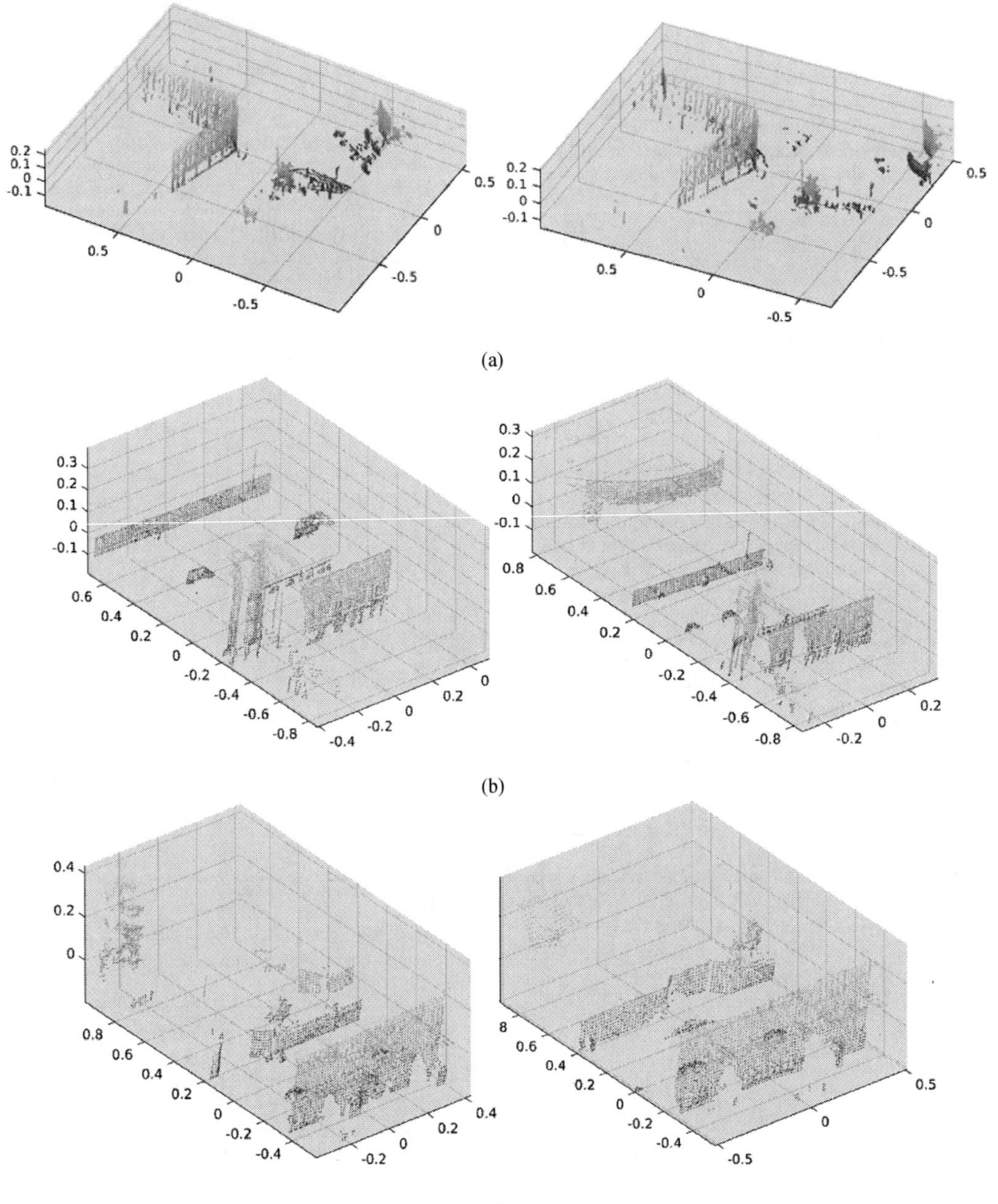

(a) (b) (c)

https://doi.org/10.5194/isprs-archives-XLIV-M-2-2020-117-2020

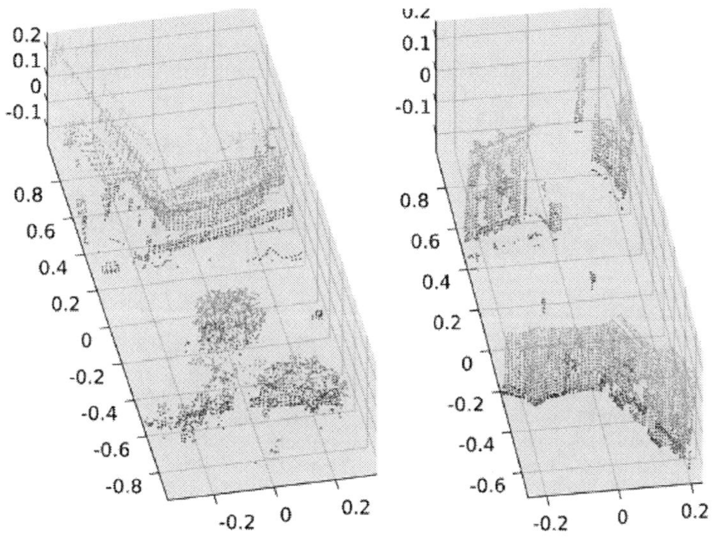

(d)

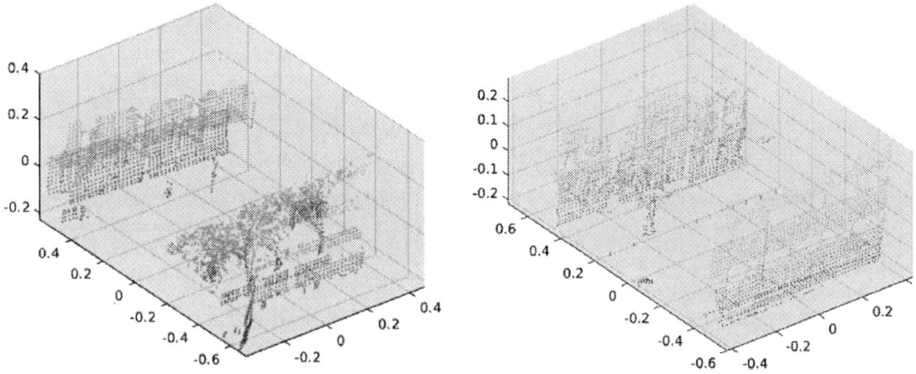

(e)

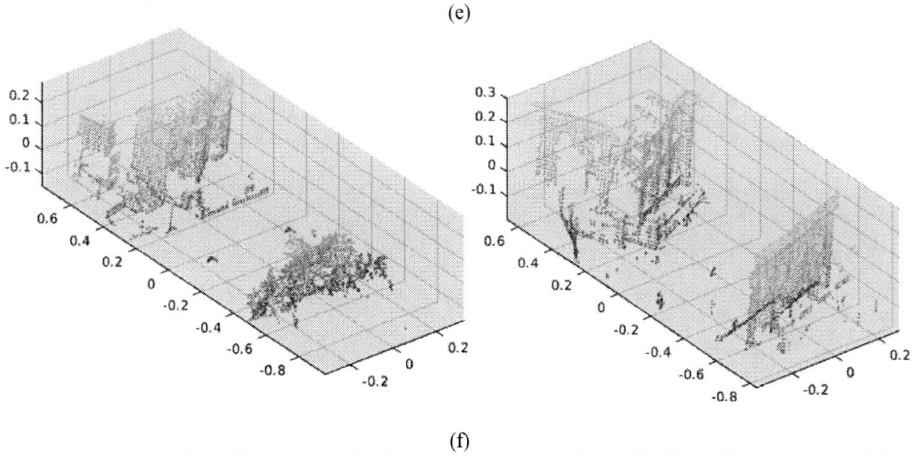

(f)

Figure 6. Examples of matching results, (a)~(c) are true positive matching, (d)~(f) are false positive matching.

REFERENCES

Angelina Uy, M. & Hee Lee, G. 2018, "Pointnetvlad: Deep point cloud based retrieval for large-scale place recognition", Proceedings of the IEEE Conference on Computer Vision and Pattern Recognition, pp. 4470.

Bosse, M. & Zlot, R. 2013, "Place recognition using keypoint voting in large 3D lidar datasets", *2013 IEEE International Conference on Robotics and Automation* IEEE, pp. 2677.

Cramariuc, A., Dubé, R., Sommer, H., Siegwart, R. & Gilitschenski, I. 2018, "Learning 3d segment descriptors for place recognition", *arXiv preprint arXiv:1804.09270.*

Dhiman, N.K., Deodhare, D. & Khemani, D. 2015, "Where am I? Creating spatial awareness in unmanned ground robots using SLAM: A survey", *Sadhana,* vol. 40, no. 5, pp. 1385-1433.

Dubé, R., Cramariuc, A., Dugas, D., Nieto, J., Siegwart, R. & Cadena, C. 2018, "SegMap: 3d segment mapping using data-driven descriptors", *arXiv preprint arXiv:1804.09557.*

Dubé, R., Dugas, D., Stumm, E., Nieto, J., Siegwart, R. & Cadena, C. 2017, "Segmatch: Segment based place recognition in 3d point clouds", *2017 IEEE International Conference on Robotics and Automation (ICRA)* IEEE, pp. 5266.

Emter, T. 2012, "Integrated Multi-Sensor Fusion and SLAM for Mobile Robots", *Proceedings of the 2011 Joint Workshop of Fraunhofer IOSB and Institute for Anthropomatics, Vision and Fusion Laboratory*KIT Scientific Publishing, pp. 91.

Emter, T. & Petereit, J. 2018, "Stochastic cloning and smoothing for fusion of multiple relative and absolute measurements for localization and mapping", *2018 15th International Conference on Control, Automation, Robotics and Vision (ICARCV)*IEEE, pp. 1508.

Koch, G., Zemel, R. & Salakhutdinov, R. 2015, "Siamese neural networks for one-shot image recognition", *ICML deep learning workshop*, Anonymous Lille.

Maddern, W., Pascoe, G., Linegar, C., Newman, P., 2017. 1 year, 1000 km: The Oxford RobotCar dataset. *The International Journal of Robotics Research* 36 (1), 3-15.

Olson, E. 2009a, "Recognizing places using spectrally clustered local matches", *Robotics and Autonomous Systems,* vol. 57, no. 12, pp. 1157-1172.

Olson, E.B. 2009b, "Real-time correlative scan matching", *2009 IEEE International Conference on Robotics and Automation*IEEE, pp. 4387.

Pedregosa, F., Varoquaux, G., Gramfort, A., Michel, V., Thirion, B., Grisel, O., Blondel, M., Prettenhofer, P., Weiss, R. & Dubourg, V. 2011, "Scikit-learn: Machine learning in Python", *the Journal of machine Learning research,* vol. 12, pp. 2825-2830.

Qi, C.R., Su, H., Nießner, M., Dai, A., Yan, M. & Guibas, L.J. 2016, "Volumetric and multi-view cnns for object classification on 3d data", *Proceedings of the IEEE conference on computer vision and pattern recognition*, pp. 5648.

Singandhupe, A. & La, H. 2019, "A review of slam techniques and security in autonomous driving", 2019 Third IEEE International Conference on Robotic Computing (IRC)IEEE, pp. 602.

Su, H., Maji, S., Kalogerakis, E. & Learned-Miller, E. 2015, "Multi-view convolutional neural networks for 3d shape recognition", *Proceedings of the IEEE international conference on computer vision*, pp. 945.

Wu, Q., Sun, K., Zhang, W., Huang, C. & Wu, X. 2016, "Visual and LiDAR-based for the mobile 3D mapping", *2016 IEEE International Conference on Robotics and Biomimetics (ROBIO)*IEEE, pp. 1522.

Yin, H., Tang, L., Ding, X., Wang, Y. & Xiong, R. 2018, "LocNet: Global localization in 3D point clouds for mobile vehicles", *2018 IEEE Intelligent Vehicles Symposium (IV)*IEEE, pp. 728.

Yin, H., Wang, Y., Ding, X., Tang, L., Huang, S. & Xiong, R. 2019, "3d lidar-based global localization using siamese neural network", *IEEE Transactions on Intelligent Transportation Systems,* vol. 21, no. 4, pp. 1380-1392.

Yin, P., He, Y., Xu, L., Peng, Y., Han, J. & Xu, W. 2018a, "Synchronous adversarial feature learning for lidar based loop closure detection", *2018 Annual American Control Conference (ACC)*IEEE, pp. 234.

Yin, P., Xu, L., Liu, Z., Li, L., Salman, H., He, Y., Xu, W., Wang, H. & Choset, H. 2018b, "Stabilize an unsupervised feature learning for LiDAR-based place recognition", *2018 IEEE/RSJ International Conference on Intelligent Robots and Systems (IROS)*IEEE, , pp. 1162.

Yin, W. & Schütze, H. 2015, "Convolutional neural network for paraphrase identification", *Proceedings of the 2015 Conference of the North American Chapter of the Association for Computational Linguistics: Human Language Technologies*, pp.901.

This contribution has been peer-reviewed.
https://doi.org/10.5194/isprs-archives-XLIV-M-2-2020-117-2020

AUTHOR INDEX

International Society for Photogrammetry and Remote Sensing (ISPRS)
c/o Leibniz University Hannover
Institute of Photogrammetry and GeoInformation
Nienburger Str. 1
D-30167 Hannover, Germany

ISBN 978-1-7138-2043-7

CURRAN ASSOCIATES INC.
proceedings.com

Eleventh International Conference on Recirculating Aquaculture 2016

Roanoke, Virginia, USA
19-21 August 2016

RASTECH